Engineering Materials

More information about this series at http://www.springer.com/series/4288

Mohd Sapuan Salit

Tropical Natural Fibre Composites

Properties, Manufacture and Applications

Springer

Mohd Sapuan Salit
Department of Mechanical and
  Manufacturing Engineering
Universiti Putra Malaysia
Serdang, Selangor
Malaysia

ISSN 1612-1317 ISSN 1868-1212 (electronic)
ISBN 978-981-287-154-1 ISBN 978-981-287-155-8 (eBook)
DOI 10.1007/978-981-287-155-8

Library of Congress Control Number: 2014944330

Springer Singapore Heidelberg New York Dordrecht London

© Springer Science+Business Media Singapore 2014
This work is subject to copyright. All rights are reserved by the Publisher, whether the whole or part of the material is concerned, specifically the rights of translation, reprinting, reuse of illustrations, recitation, broadcasting, reproduction on microfilms or in any other physical way, and transmission or information storage and retrieval, electronic adaptation, computer software, or by similar or dissimilar methodology now known or hereafter developed. Exempted from this legal reservation are brief excerpts in connection with reviews or scholarly analysis or material supplied specifically for the purpose of being entered and executed on a computer system, for exclusive use by the purchaser of the work. Duplication of this publication or parts thereof is permitted only under the provisions of the Copyright Law of the Publisher's location, in its current version, and permission for use must always be obtained from Springer. Permissions for use may be obtained through RightsLink at the Copyright Clearance Center. Violations are liable to prosecution under the respective Copyright Law. The use of general descriptive names, registered names, trademarks, service marks, etc. in this publication does not imply, even in the absence of a specific statement, that such names are exempt from the relevant protective laws and regulations and therefore free for general use.
While the advice and information in this book are believed to be true and accurate at the date of publication, neither the authors nor the editors nor the publisher can accept any legal responsibility for any errors or omissions that may be made. The publisher makes no warranty, express or implied, with respect to the material contained herein.

Printed on acid-free paper

Springer is part of Springer Science+Business Media (www.springer.com)

Preface

In tropical regions, natural fibres are found in abundance and many of these fibres are suitable to be used as fibres in composites. The work on natural fibre composites had been carried for many centuries. Ford had used biopolymer from soya and filled with some natural fibres in their experimental cars in the 1950s. Nowadays, the research and development of natural fibre composites have been intensified and many industries have reaped the benefits from their development. As far as the use of fibres is concerned, fibres such as kenaf, oil palm, jute, henequen, sisal, hemp, banana stem, abaca, pineapple leaf, flax and bagasse are suitable candidates, but in this book only fibres that found in abundance in tropical regions are reported. Natural fibres have many advantages that made them suitable to be used as reinforcement or filler in some polymer matrices such as low cost, renewable, abundance, posing no environmental hazard, ease of handling, acceptable specific strength and stiffness properties and aesthetically pleasing. Many research works in the past have been carried out on the characterisation and determination of important properties of natural fibres and their composites. To some extent, natural fibre composites have been used in many products and components such as furniture, automotive components, building and construction industries, appliances and marine industry. However, its use is only limited to mainly non and semi-structural components. The call for green materials has generated interests to use natural fibres with bio-based polymers and fully biodegradable materials can be developed. Attempts are made to produce natural fibre composites with the matrices made from commercially available biopolymers or bio-polymers developed in house. This book is concerned with tropical natural fibres and their composites with polymer materials. However, it focuses on tropical natural fibres and their composites as books on natural fibre composites are now growing in number. It concentrates mainly on selected tropical natural fibres namely kenaf, oil palm, pineapple leaf, sugar palm, banana stem, sugarcane and coconut fibres. In addition, this book presents some new aspects related to natural fibre composites such as design, materials selection and manufacturing process.

This book comprises seven chapters. Chapter 1 provides general information about composite materials. Important aspects of tropical natural fibres are presented in Chap. 2, which include advantages and disadvantages of natural fibres, description of seven selected tropical natural fibres, and typical applications of tropical natural fibres. The drive for 'green' materials has triggered the need for inclusion of a chapter titled biopolymers. Biopolymers can be used as matrices for natural fibres along with synthetic polymers. In Chap. 4, mechanical properties, which include tensile, flexural and impact properties of natural fibre composites are presented. New topic on the design of tropical natural fibre composite products is given in Chap. 5. Possible manufacturing techniques for natural fibre composites are included in Chap. 6. Finally, a brief chapter on selected products developed from tropical natural fibre composites can be found in Chap. 7.

I am indebted to various organisations and individuals who have made writing of this book possible. The permission and financial allowance given by my employer Universiti Putra Malaysia are highly appreciated that enabled me to carry out this task during my sabbatical leave from October 2013 to June 2014. I am indebted to the cooperation of Dr. Nukman Yusuff of Universiti Malaya who provided facility during my sabbatical leave. I would like to thank my family member; my wife Ms. Nadiah Zainal Abidan, and my daughter Qurratu Aini who have endured the pains and difficulties in the course of preparation of this book. I wish to thank my research assistant, Mr. Wan Mohamad Haniffah Wan Hussin for his assistance.

Mohd Sapuan Salit

Contents

1 Introduction 1
  1.1 Background 1
  1.2 History of Composites. 4
  1.3 Advantages of Composites. 5
  1.4 Disadvantages of Composites. 6
  1.5 Classification of Composites 7
    1.5.1 Polymer Matrix Composites (PMC) 7
    1.5.2 Metal Matrix Composites (MMC). 7
    1.5.3 Ceramic Matrix Composites (CMC) 9
  1.6 Advanced Composites. 10
  1.7 Nanocomposites 11
  1.8 Natural Fibre Composites 11
  1.9 Tropical Natural Fibre Reinforced Polymer Composites 12
  References 14

2 Tropical Natural Fibres and Their Properties 15
  2.1 Background 15
  2.2 Advantages of Natural Fibres. 16
  2.3 Disadvantages of Natural Fibres. 17
  2.4 Types of Natural Fibres. 17
    2.4.1 Banana Fibres 17
    2.4.2 Coconut Fibres. 20
    2.4.3 Kenaf Fibres 24
    2.4.4 Oil Palm Fibres 27
    2.4.5 Sugar Palm Fibres 29
    2.4.6 Sugarcane Fibres 31
    2.4.7 Pineapple Leaf Fibres 33
  References 36

vii

3	**Biopolymer**		39
	3.1	Introduction	39
	3.2	Classification of Biopolymers	40
		3.2.1 Polylactide (PLA)	40
		3.2.2 Thermoplastic Starch (TPS)	41
		3.2.3 Cellulose	45
		3.2.4 Polyhydroxyalkanoates (PHAs)	47
		3.2.5 Synthetic Biopolymer	48
	3.3	Conclusions	49
	References		50

4	**Mechanical and Other Related Properties of Tropical Natural Fibre Composites**		53
	4.1	Introduction	53
	4.2	Banana Fibre Composites	54
	4.3	Coconut Fibre Composites	56
	4.4	Kenaf Fibre Composites	57
	4.5	Oil Palm Fibre Composites	60
	4.6	Sugar Palm Fibre Composites	61
	4.7	Sugarcane Fibre Composites	64
	4.8	Pineapple Fibre Composites	66
	4.9	Fibre-Matrix Interface in Tropical Natural Fibre Composites	69
	4.10	Water/Moisture Absorption	70
	References		70

5	**Design and Materials Selection of Tropical Natural Fibre Composites**		75
	5.1	Introduction	75
	5.2	Engineering Design Process	76
	5.3	Conceptual Design	76
		5.3.1 Market Investigation	77
		5.3.2 Product Design Specification (PDS)	77
		5.3.3 Concept Generation	81
		5.3.4 Concept Evaluation	84
	5.4	Detail Design	88
	5.5	Design for Manufacture	90
	5.6	Materials Selection in Design	92
	5.7	Materials Selection for Tropical Natural Fibre Composites	95
	References		99

6 Manufacturing Techniques of Tropical Natural Fibre Composites

6	**Manufacturing Techniques of Tropical Natural Fibre Composites**	103
	6.1 Introduction	103
	6.2 Hand Lay-up	104
	6.3 Filament Winding	105
	6.4 Resin Transfer Moulding	108
	6.5 Pultrusion	108
	6.6 Compression Moulding	110
	6.7 Injection Moulding	113
	6.8 Vacuum Bag Moulding	115
	6.9 Extrusion	115
	6.10 Vacuum Assisted Resin Infusion (VARI)	116
	References	117
7	**Applications of Tropical Natural Fibre Composites**	119
	7.1 Introduction	119
	7.2 Application	119
	7.3 Miscellaneous Products Developed by the Author	121
	References	123

Chapter 1
Introduction

Abstract In this chapter, background of composite materials is presented followed by discussion of various major classifications of composites such as polymer matrix composites, metal matrix composites and ceramic matrix composites. Sections of history of composites, advantages and disadvantages of composites, advanced composites, nanocomposites and natural fibre composites are made available to the readers. This chapter is presented as background information for the topic of the main concern in this book, i.e. tropical natural fibre reinforced composites. Tropical natural fibre composites are chosen because majority of the contents of this book are based on the work carried out on natural fibre composites in one of the tropical country, i.e. Malaysia.

Keywords Composites · Polymer composites · Tropical natural fibre composites · Metal matrix composites · Ceramic matrix composites

1.1 Background

The competition of technology among nations is to create better standard of living for the benefit of mankind. The products in the marketplace should be developed by considering the customer interests and the products should be developed at low cost, high quality and in the minimum possible time [13]. Therefore, the mankind is seeking for new product, material and technology to serve this purpose. Materials play an important role in the advancement of human life. In the past, the influence of materials can be felt so much so that an era was named after the materials like stone age, bronze age, and iron age. But, nowadays, it is difficult to associate this era with a particular type of material and if it can be given such name, it is more appropriate to call it the age of various materials. Polymers, metals, and ceramics, are the most common monolithic materials and the composites with those materials as matrices are mainly available. For the future, smart materials, metamaterials and nano materials will be among the major contenders

---

© Springer Science+Business Media Singapore 2014
M.S. Salit, *Tropical Natural Fibre Composites*,
Engineering Materials, DOI 10.1007/978-981-287-155-8_1

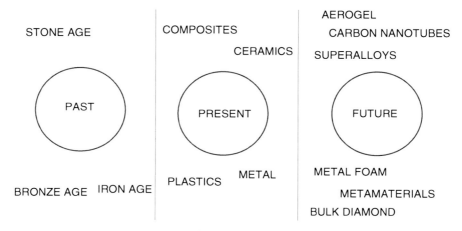

Fig. 1.1 Materials in the past, present and future

(see Fig. 1.1). In fact research on nano materials has been intensified in the recent years and some of the research works have led to successful commercialization.

Composite materials or composites are one of the major developments in material technology in the recent years. Composite materials have made significant impact in engineering and technology fields recently as they demonstrate combined properties of reinforcement and matrix to form stronger and stiffer materials that cannot be obtained from the constituent materials acting alone. In addition, their light weight attribute has made them even more attractive. It was the reason why Easterling [3] had called them as 'selected wonders of the world of advanced materials'. Components made from composites are normally found in automotive, consumer goods, aerospace, building, textile, furniture, medical, railway, defense, marine, sport, and electronic industries. Figure 1.2 shows jet skies made from composites.

Composites are playing an ever increasing role in engineering applications. They are used in domestic articles and components in aircraft and automobiles. Materials manufacturers are trying constantly to improve and develop these types of materials. Suppliers of these materials are endevouring to provide materials that meet customers' requirements and expectation. This is the principal reason for their rapid progress in terms of utilization and product diversification.

Composites are a combination of two or more constituent materials where the performance of composites is better that their constituent materials acting alone. In composites a distinct reinforcement is distributed in a matrix and the resultant composites generally have physical and chemical properties superior to and different from reinforcement or matrix. But at macroscopic level (in final composite product) constituent materials retain their properties.

The various types of composites available include particles in matrix, and fibres in matrix or a combination of them. The manufacturing of composites provide a means of improving stiffness, strength, resistance to temperature and creep of composites whilst at the same time offering cost reduction. In these materials, the

1.1 Background

Fig. 1.2 Jet skies made from composites

strength and stiffness of load bearing fibres are married to matrices to provide a good combination of strength and toughness. In polymer composites, the fibres are normally glass, carbon and aramid. Aramid fibres are textile fibres, spun from an organic polymer. The most famous aramid fibre used commercially is Kevlar® developed by DuPont in 1973.

The fibres are very brittle and alone their strength and stiffness are not fully realized. Fibres can be arranged in the matrix depending on the type fibres and the final properties of composites [5]. Fibres are normally arranged in continuous form either unidirectionally, bidirectionally or multidirectionally [9]. Fibres can also be arranged in discontinuous form or sometimes called short fibres either unidirectionally or arranged randomly. In many cases, fibres are made in the form of particles and these are normally called filled or particulate composites. The fibres could also be in the form of chopped strand mat. Figure 1.3 shows continuous glass fibre yarn. The matrix protects the fibres and transfers the load to them. There are two types of polymer used for matrix; thermoplastic and thermosetting.

Thermoplastic materials can be heated and formed repeatedly and normally used for injection moulding applications. They are not cross-linked and they obtained strength and stiffness from the properties of the monomer units and the degree of entanglement of the polymer chains [8]. These include acrylics, nylons, polypropylenes, polycarbonates, polystyrenes, poly(vinyl chlorides) (PVC), acrylonitrile/butadiene/styrene (ABS), polytetrafluoroethylene (PTFE), polyethylene terephthalate (PET), polybutynene terephthalate (PBT), polyphenylene sulfide (PPS) and polysulfone (PS). Thermosetting materials are formed through an initial heating and cure and cannot be melted and reformed. During curing, they formed three-dimensional molecular chains, called cross-linking. These materials include polyesters, melamine, phenolics, cyanate esters, bismaleimide (BMI), polyimide, polyurethane, polyester, vinylesters and epoxides.

Fig. 1.3 Continuous glass fibre roving

Thermoplastics offer a wide range of advantages over thermosetting matrices, including unlimited shelf life, short process cycle times, their ability to be reformed, increased moisture resistance, increased toughness and impact resistance. However, there are several disadvantages associated with thermoplastic matrix composites. The present cost of the thermoplastic materials is higher than that of for conventional thermosetting materials.

1.2 History of Composites

Herakovich [6] reported that fibre composites from the papyrus plant were used for writing materials in Egypt in 4000 B.C. Historically straws were used as reinforcement for mud and clay to form bricks for building construction. The process of brick-making was found on ancient Egyptian tomb painting. However, the use of composites has been reported as early as the reign of Pharaoh in the Ancient Egypt where the bricks were reinforced with straw. During that time, the composite bricks were 'cured' in the sun and the bricks were not fired in the oven because there was no rain [4]. Gordon [4] also reported that laminated composites were developed in 1920 from cellulose paper or fabric being impregnated with phenolic and was available for machining into cams and gears and bearings.

However, engineered composites were reported to be introduced in 1940s where glass fibre was used as reinforcement of polymers in aerospace, marine and automotive industries. Polymers are the most widely used matrix material for composites. Composites have been used since 1940s as materials for aerospace, automotive and building industries. The main reason for using these materials is due to their light weight. During that period, majority of resins used include epoxy and polyester. Murphy [11] reported that in 1941, glass cloth fibre reinforced allyl composites were used in the USA. Glass fibre reinforced polyester composites were developed in 1942 by Owen-Corning for the World War Two aircraft [2]. Later in 1960s carbon fibres were introduced as reinforcement for composites particularly in aerospace and sporting goods. In these applications, the prime concern was performance rather than cost and this was the beginning of the development of advanced composites.

1.3 Advantages of Composites

Composite materials have been accepted as structural, semi and non-structural applications in many industries and they still have bright future in more demanding applications. There are numerous advantages in using composites and they listed as follows:

- High strength-to-weight or modulus-to-weight ratio
- Reduced weight (low density)
- Properties that can be tailor-made with each composite type
- Very good creep and fatigue performances and carbon fibre composites virtually fatigue free
- Longer life (no corrosion)
- Net-shape or near-net-shape parts can be produced
- Low thermal insulation properties
- Lower manufacturing costs—lower part count
- Inherently high internal damping
- Concurrent engineering (design for manufacture) concept can be used
- Can be design to give desired thermal expansion
- Increased (or decreased) thermal and/or electrical conductivity
- Can be used in the presence of water, process liquids, oil or grease as lubricant
- Impermeable to chemicals, gases, liquids
- Complex shapes can be manufactured at lower cost compared to fabricated or machined metal alloys
- High toughness
- Good dimensional stability
- Good wear resistance under heavy load
- Generally electrical insulating
- Non magnetic
- Relatively easy to use
- Good shelf life at room temperature
- Can be manufactured in a variety of sizes suitable for a range of components
- High integral strength
- High impact resistance

- Fibres can be oriented in the direction of the principal stresses to give high structural efficiency
- Some structured can be repaired
- Production equipment generally operates at lower pressures than with metals
- Good shelf life at room temperature
- Provide reductions in cost of tooling, making prototype and fabrication
- Improved damaged tolerance
- Aesthetically pleasing
- Better noise, vibration and harshness (NVH) characteristics compared to metals counterpart
- Lower thermal expansion than metals.

1.4 Disadvantages of Composites

However, since composite materials technology is still not matured yet, the composites also suffered from various drawbacks such as:

- High cost of raw materials and fabrication
- Properties can be anisotropic
- Possible weakness of transverse properties
- Weak matrix and low toughness
- Environmental degradation of matrix
- Difficulty in attaching
- Susceptibility to splitting and delamination
- Cracks often propagate along fibre direction
- Sometimes degraded by water and solvents
- Graphite fibre materials can cause corrosion of adjacent materials
- Difficulty with analysis
- Residual stresses due to shrinkage during curing
- Inherent variation in component properties due to the nature of fabrication technique
- Moisture absorption affects properties
- Inconsistent design and materials properties data
- Stiffness of glass fibre composites is lower than metals
- Special storage is required
- Majority of them cannot be recycled
- High labour cost (hand lay-up)
- Training can be expansive (hand lay-up)
- Pose health and safety problems
- Design and manufacturing technologies are not established
- Composite definition is very broad—from nanocomposites to metal matrix composites

1.4 Disadvantages of Composites

- Despite composite technology is not established yet, research on nanocomposites is given too much attention
- Moisture absorption is high and it affects dimensional stability

1.5 Classification of Composites

Composites can be classified based on the type of matrices used and they can be a polymer, a metal or a ceramic.

1.5.1 Polymer Matrix Composites (PMC)

Polymer matrix composites are highly developed composites and polymer either thermoplastic or thermoset was traditionally used as matrix in modern composites. Thermoset is a polymer that cannot return to its original solid state after being heated while thermoplastic is a polymer that can be softened upon processing but it returns to solid state after processing is complete. Sometimes, as far as classification of polymer is concerned, rubber is considered as another category of polymer. PMC had been the materials of interests in material field and they have been around for almost 70 years. MMC and CMC are considered as new materials being introduced around two decades ago. PMC had been used in many industries such as automotive, aerospace, marine, building, furniture, electronic, oil and gas and sport and leisure industries. The most common manufacturing techniques involving PMC include hand lay-up, spray up, pultrusion, filament winding, compression moulding, injection moulding, prepreg moulding, autoclave moulding, resin transfer moulding, calendaring, braiding, thermoforming and vacuum bag moulding. Some of the information on PMC have been discussed in earlier sections. Figure 1.4 shows Injection moulded glass fibre reinforced nylon composite clutch pedal while Fig. 1.5 shows the pultrusion process of glass fibre reinforced epoxy composite rods. Drive shafts made from filament wound carbon fibre reinforced epoxy composites are shown in Fig. 1.6. Figure 1.7 shows a bench made from pultruded composites and Fig. 1.8 shows a water filter made from filament winding process.

1.5.2 Metal Matrix Composites (MMC)

Metal matrix composites are composite materials made from soft and light metal matrices reinforced with fibres, hard particulate, platelets, whiskers, and metal wires [14]. The matrix for MMC is normally metal alloys, rather than pure metals. The major advantages of MMC include wear resistance, thermal conductivity, high

Fig. 1.4 Injection moulded glass fibre reinforced nylon composite clutch pedal

Fig. 1.5 Pultrusion process of glass fibre reinforced epoxy composite rods

specific strength and stiffness, enhanced hardness and toughness properties, low coefficient of thermal expansion, good creep resistance and prolonging fatigue life. It can be found mainly in automotive and aerospace industries. Products from metal matrix composites can be fabricated using solid state processing (diffusion bonding, hot isostatic pressing (HIP) and powder metallurgy), liquid state processing (melt stirring, stir casting, spray deposition, compocasting or rheocasting, squeeze casting, and liquid melt infiltration), co-deposition, and in situ processes [10].

1.5 Classification of Composites

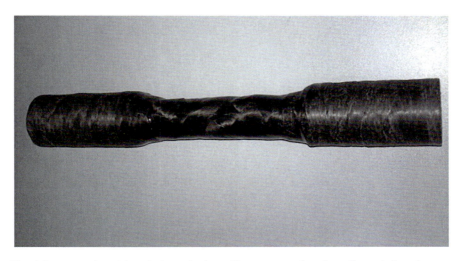

Fig. 1.6 Automotive drive shaft made from filament wound carbon fibre reinforced epoxy composites

Fig. 1.7 A bench made from pultruded glass fibre composites

1.5.3 Ceramic Matrix Composites (CMC)

Ceramic matrix composites are composites made from ceramic matrix and reinforced by fibres or whiskers. Ceramic was normally regarded as a brittle material and ceramic matrix composite was introduced to increase the toughness of ceramic [10]. Schwartz [14] has enumerated the advantages of CMC as low density, high

Fig. 1.8 Water filter made from filament wound composites

hardness, high specific strength and stiffness, high fracture toughness, and the ability to tailor properties for end-use specification. Typical manufacturing processes to produce components from CMC include conventional mixing and pressing, techniques involving slurries, liquid state processing (matrix transfer moulding, and pyrolysis of polymer), sol-gel processing, vapour deposition techniques, Lanxide process, and in situ techniques. Details about these processes can found in Matthews and Rawlings [10].

1.6 Advanced Composites

If the fibres used in composites are carbon and Kevlar® fibres, then the composites are normally referred to as advanced composites. Compare to conventional composites like glass fibre composites, advanced composites have high specific strength and stiffness properties, more expansive to produce and normally used in advanced applications like in aerospace and defense industries where tailor made composites are normally targeted [2]. Apart from these advanced composites that are normally used in advanced applications like in aerospace and defense industries, other composites are regarded as engineering composites and the most popular among them is used in laymen term as fiberglass. In actual fact, engineering composites can also employ other types of fibres like natural fibres or fillers.

1.7 Nanocomposites

Nanocomposites are emerging materials where their technology is still under development. Nanocomposites are composites in which the reinforcement has the dimension in nanometer (10^{-9} m) [9]. Nanocomposites, just like conventional composites, are combination of two or more different constituent materials which after being mixed the individual constituent retains its properties but the composites give the optimum properties. The major difference of nanocomposites from conventional composites is that the filler material is in the nanoscale dimension normally less than 100 nm. Nano fillers are such amazing materials as by adding a small amount in polymer, they would enhance the properties of nanocomposites such good scratch resistance, heat stability, chemical resistance, elasticity, and conductivity. Carbon nanotube (CNT), clay, silica, silver nanowire and graphane are among the nanoscale filler materials used in nanocomposites. Nanocomposites can be found in automotive, biomedical, sensor, transistor, optics, electronics, optoelectronics, energy storage, solar cell, smart textile applications.

1.8 Natural Fibre Composites

Synthetic fibres are expansive, non-biodegradable and pollute the environment. With this in view research on natural fibres were intensified as they have huge potential to replace some synthetic fibres. According to Westman et al. [15] research and development on the use of natural fiber reinforced composites has been in existence as early as 1900s. It has only attracted the interests of many researchers until late in the 1980s. Back in 1930s Henry Ford developed soya based matrix reinforced with natural fibres to form natural fibre composites used in car body panel [7]. Such development was driven by economic reason. But nowadays, the interest in the use of natural fibre reinforced plant based polymer composites is mainly due to the environmental reason.

Kenaf, sisal, jute, hemp, abaca, oil palm, pineapple leaf, banana stem, bagasse, and coir are the natural fibres reported to be used as reinforcements or fillers in polymer composites. Natural fibre composites find their applications in many non and semi-structural components such as in furniture industry, building materials, and automotive components due to some advantages they offer. Light weight, corrosion resistance, abundantly available, renewable, cheap and comparable mechanical properties are among the most desirable properties that trigger their use.

Fig. 1.9 Photos and micrographs of durian skin (**a**) and durian skin fibre (**b–d**) (Anuar and Sapuan [1]; Reproduced with permission)

Natural fibre composites have found their application in many engineering products. The benefits of natural fibre composites include [16]:

- Low volumetric cost
- Improvement of 'wood' surface appearance
- Increase of stiffness of thermoplastics
- Increase of heat deflection temperature

1.9 Tropical Natural Fibre Reinforced Polymer Composites

The tropic is a region of the Earth surrounding the Equator. 'Tropical natural fibres' are the fibres can be obtained from the region. Malaysia is located in the tropics. This book is focusing on the use of natural fibres mainly obtained in

1.9 Tropical Natural Fibre Reinforced Polymer Composites

Fig. 1.10 Non woven betel nut fibres

Malaysia and in other tropical countries. Fibres obtained from Malaysia are mainly discussed in this book and those fibres include oil palm, sugar palm, sugar cane, banana stem, pineapple leaf, kenaf and coconut fibres. Although kenaf is not a native tree of Malaysia, nowadays, it can be found in abundance because kenaf has been identified by the Malaysian government as an important commodity in Malaysia. A government agency called National Kenaf and Tobacco Board had been established recently to carry out research, development, monitoring and commercialization of kenaf and kenaf composite products in Malaysia [12]. Figure 1.9 shows the photos and scanning electron micrographs of durian skins and their fibres. Durian is a tropical fruit normally grown in tropical regions like Malaysia, Thailand and Indonesia and in Malaysia, durian is known as 'the king of fruits'. The author and his co-researchers have embarked on the research of durian skin composites as alternative materials to the traditional composites. Figure 1.10 shows betel nut fibre; a source of fibre for tropical natural fibre composites. The work on tropical betel nut fibre composites is recently being initiated by the author.

References

1. Anuar, H., Sapuan, S.M.: Durian skin fibre and its composites. In: Proceedings of Postgraduate Symposium on Composites Science and Technology 2014, 28 Jan, pp. 1–4. Putrajaya, Malaysia (2014)
2. Dorworth, L.C., Gardiner, G.L., Mellema, G.M.: Essentials of Advanced Composite Fabrication and Repair. Aviation Supplies & Academics Inc, Newcastle (2009)
3. Easterling, K.: Tomorrow's Materials, 2nd edn. The Institute of Materials, London (1990)
4. Gordon, J.E.: The New Science of Strong Materials or Why You Don't Fall through the Floor. Princeton University Press, Princeton (2006)
5. Hancox, N.L., Sajic, P., Tattersall, P., Tetlow, R., Lovell, D.R.: A Guide to High Performance Plastics Composites. British Plastics Federation, London (1980)
6. Herakovich, C.T.: Mechanics of Fibrous Composites. Wiley, New York (1998)
7. Hogg, P.J.: The role of materials in creating a sustainable economy, Plenary Paper, Fourth International Materials Technology Conference and Exhibition, 23rd–26th March, Kuala Lumpur, Malaysia (2004)
8. Hyer, M.W.: Stress Analysis of Fiber-Reinforced Composite Materials. WCB McGraw-Hill, Boston (1998)
9. Mallick, P.K.: Fiber-Reinforced Composites Materials, Manufacturing, and Design, 3rd edn. CRC Press, Boca Raton (2008)
10. Matthews, F.L., Rawlings, R.D.: Composite Materials: Engineering and Science. Chapman & Hall, London (1994)
11. Murphy, J.: Reinforced Plastics Handbook. Elsevier Advanced Technology, Oxford (1994)
12. Paridah, M.T., Shukur, N.A.A., Harun, J., Abdan, K.: Overview: Kenaf—A journey towards enegizing the biocomposite industry in Malaysia. In: Paridah, M.T., Abdullah, L.C., Kamaruddin, N., (eds.) Kenaf: Biocomposites, Derivatives and Economics, Pustaka Prinsip Sdn. Bhd., Kuala Lumpur (2009)
13. Sapuan, S.M.: Concurrent Engineering for Composites. UPM Press, Serdang (2010)
14. Schwartz, M.M.: Composite Materials, Volume I: Properties, Nondestructive Testing, and Repair. Prentice Hall PTR., Upper Saddle River (1997)
15. Westman, M.P., Laddha, S.G., Fifield, L.S., Kafentzis, T.A., Simmons, K.L.: Natural Fiber Composites: A Review, report no. PNNL-19220. Report prepared for U.S. Department of Energy (2010)
16. Zheng, Y., Cao, D., Wang, D., Chen, J.: Study on the interface modification of bagasse fibre and the mechanical properties of its composite with PVC. Composites A **38**, 20–25 (2007)

Chapter 2
Tropical Natural Fibres and Their Properties

Abstract In this chapter, a background of the importance of natural fibres is presented. The advantages and disadvantages of tropical natural fibres are listed. The chapter elaborates seven types of tropical natural fibres commonly being studied and used. The information about fibre extraction process, the application of fibres and other important topics are discussed.

Keywords Fibre extraction · Advantages of natural fibres · Natural fibre products · Tropical natural fibres · Kenaf fibres

2.1 Background

Natural fibre encompasses all forms of fibres from woody plants, grasses, fruits, agriculture crops, seeds, water plants, palms, wild plants, leaves, animal feathers, and animal skins. By-products of pineapple, banana, rice, sugarcane, coconut, oil palm, kenaf, hemp, cotton, abaca, sugar palm, sisal, jute and bamboo are among the fibres known to be used to make composites. Wool and silk are strong fibrous materials and wool had been used in textile industry dated back from 35,000 years ago and silk from at least 5,000 years [38]. The ancient Egyptians had been reported to have used natural fibre composites, made from straw and clay or mud around 3,000 years ago. But in this book, the study is restricted only to plant based fibres. In the recent years, there has been a growing interest in the application of natural fibres as reinforcements for polymer matrices. Natural fibre has good potential as reinforcement in thermoplastic and thermoset polymer composites mainly due to low density and high specific properties of fibres. Natural fibres have the properties, composition, structures and features that are suitable to be used as reinforcements or fillers in polymer composites.

The plant based fibres contain cellulose and non-cellulose materials such as hemicelluloses, pectin and lignin; thus they are also known as lignocellulosic or cellulosic fibres. Cellulose is semicrystalline polysaccharide found in natural fibres

© Springer Science+Business Media Singapore 2014
M.S. Salit, *Tropical Natural Fibre Composites*,
Engineering Materials, DOI 10.1007/978-981-287-155-8_2

15

and it is the reason for the natural fibres to demonstrate hydrophilic behaviour. It provides strength and rigidity to the fibres. Hemicellulose is an amorphous polysaccharide and its molecular weight is lower than cellulose. Fibres are held together by means of pectin. Pectin is a class of plant cell wall polysaccharide that can be found in a plants' cell wall. Lignin acts as a binder for the cellulose fibres and it adds strength and stiffness to the cell walls. Holocellulose contains mainly of *cellulose* and hemicelluloses and it is the total polysaccharide of natural fibres. It is obtained after removal of extractives and lignin from natural fibres. A Lumen is a cavity inside fibre cells. Ash and wax are normally contained in the fibers.

Natural cellulose fibres are extracted from lignocellulosic by-products using biological retting (bacteria and fungi), chemical retting (boiling in chemicals), mechanical retting (hammering, decortications), and water retting. Natural fibres can be used in the form of particulate or filler, short fibres, long fibres, continuous roving, woven fabric and non-woven fabric.

2.2 Advantages of Natural Fibres

Comparing to conventional reinforcing fibres like glass, carbon and Kevlar®, natural fibres have the following advantages:

- Environmentally friendly
- Fully biodegradable
- Non toxic
- Easy to handle
- Non abrasive during processing and use
- Low density/light weight
- Compostable
- Source of income for rural/agricultural community
- Good insulation against heat and noise [24]
- Renewable, abundant and continuous supply of raw materials
- Low cost
- Enhanced energy recovery
- Free from health hazard (cause no skin irritations)
- Acceptable specific strength properties
- High toughness
- Good thermal properties
- Reduced tool wear
- Reduced dermal and respiratory irritation
- Ease of separation
- The abrasive nature of natural fibres is much lower compared to that of glass fibres, which offers advantages with respect to processing techniques and recycling [10].

2.3 Disadvantages of Natural Fibres

However, natural fibre suffered from the following drawbacks:

- Poor compatibility with hydrophobic polymer matrix
- The fibres degrade after being stored for a long period
- The inherent high moisture absorption
- The relatively high moisture absorption
- The tendency to form aggregates during processing [10]
- The low resistance to moisture [10]
- Low thermal stability [26]
- Hygroscopicity.

2.4 Types of Natural Fibres

In this section a review of various types of tropical natural fibres is made. Apart from general information about the fibres, typical applications of fibres are also presented.

2.4.1 Banana Fibres

Banana (*Musa*) is a high herbaceous plant (see Fig. 2.1) normally of 2–16 m high [6]. Although banana leaves (Fig. 2.2) were reported to be used as fibres in polymer composites, majority of work on banana fibres focused on the use of banana pseudo-stem (trunk) fibres (Fig. 2.3) as the reinforcement or filler in polymer composites. Pseudo-stem fibre is a bast fibre and it can be extracted after the fruit bunch was harvested by scrapping with a blunt knife or by using an extractor machine. Banana stem fibres are extracted by initially cutting into lengths of convenient size, and peeling layer-wise (see Fig. 2.4). The individual sheaths were dried under sun for 2 weeks (see Fig. 2.5) and then they were soaked in water for two more weeks. Once the lignin and cellulose were separated, the sheaths were dried again and the fibres were ripped off [42]. Typical density of banana fibre is 1,350 kg/m^3, cellulose/lignin ratio is 64/5, modulus is 27–32 GPa, ultimate tensile strength is 529–914 MPa and water absorption is 10–11 % [7]. Joseph et al. [15] reported that the elongation and toughness of typical banana fibre were 3.0 % and 816 MN/m^2 respectively. Banana fibre has a non-mesh structure and has long filaments [29] Fig. 2.6 shows banana pseudo-stem fibres in woven mat form.

Bilba et al. [6] studied botanical composition, thermal degradation and textural observations of banana leaf and stem before they can be proposed as reinforcements in composites. Benítez et al. [5] investigated the effect of physical and

Fig. 2.1 Banana trees

Fig. 2.2 Banana leaf

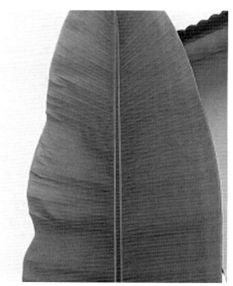

chemical treatments of banana fibres in order to use them as reinforcements in polymer composites and the specimens were prepared using injection molding process. Guimarães et al. [12] carried out studies on chemical composition, X-ray

2.4 Types of Natural Fibres

Fig. 2.3 Banana pseudo-stem

Fig. 2.4 Preparing of banana pseudo-stem fibres

powder diffraction analysis, morphological analysis and thermal behaviour of banana fibres. Thermal stability of the fibres was around 200 °C and decomposition of both cellulose and hemicelluloses in the fibres took place at 300 °C and above, while the degradation of fibres took place above 400 °C (Fig. 2.7).

Fig. 2.5 Drying of banana pseudo-stem fibres under the sun

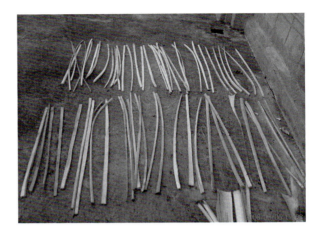

Fig. 2.6 Woven banana fibres

Banana fibres are used to make high quality textile for generations and in Japan it was used to make famous Japanese dress called kimono. It was also reported that banana fibres were used as reinforcing fibres in polymer composites and in paper making [31]. The fibre can also be used as raw material for board and cellulose derivatives.

2.4.2 Coconut Fibres

Coconut (*Cocos nucifera*) is the plant of a species of palm. It is a tropical plant (Fig. 2.8) of the Areceae (Palmae) family. Coconut fibres are mainly taken from coirs (Figs. 2.9, 2.10 and 2.11), and to a lesser extent, coconut shell (Fig. 2.12) and

2.4 Types of Natural Fibres

Fig. 2.7 Steps in preparation of banana fibres **a** banana plant **b** cutting **c** drying **d** soaking **e** re-drying **f** weaving

Fig. 2.8 A coconut plant

Fig. 2.9 Young coconut fruits

Fig. 2.10 Matured coconut fruit

spathe is used normally in the form of fillers. Coconut spathe, the covering of the coconut inflorescence, is an under-exploited material with considerable potential. This part of coconut tree is left out because demonstrates no good mechanical properties. Spathe is used as decorative (Fig. 2.13) as sold in gift shops in Kota Kinabalu, Sabah, Malaysia.

Substantial research has been carried out on coconut coir fibre and coconut shell filler and their composites. Coir is the seed-hair fibrous material found between the hard, internal shell and the outer coat (endocarp) or husk of a coconut. Coir fibre is a coarse, stiff and reddish brown fibre and is made up of smaller threads, consists of lignin, a woody plant substance, and cellulose [21]. Coir has been used for making twine, mats and brooms. It was also used in hydroponic growing [20].

2.4 Types of Natural Fibres

Fig. 2.11 Coir fibres

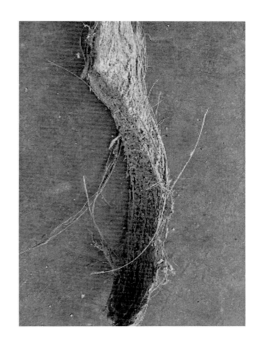

Fig. 2.12 Coconut shells

Fig. 2.13 Coconut spathe

Fig. 2.14 Kenaf plant (Courtesy of Mr. Wan Mohd Haniffah Wan Husain)

2.4.3 Kenaf Fibres

Kenaf (*Hibiscus cannabinus L.*) was a native of West Africa and had been cultivated from around 4000 B.C. [3]. It is a member of the Hibiscus gene and a family of Malvacea which is similar to cotton and okra [28]. Kenaf is a warm season plant, which requires a short period of sunlight. It has been grown for several 1,000 years for fibre and food. It is a common wild plant of tropical and subtropical Africa and Asia. It is a high carbon dioxide absorbent plant. Kenaf is a fast growing tree and could be harvested in just 4–5 months. It has very short life cycle and cultivation of kenaf produced high biomass output [35]. Kenaf stalk is made up of a soft inner core and a fibrous outer bast surrounding the core (Fig. 2.14).

Kenaf bast fibre is longer than soft wood fibre, i.e. 10 mm for the former [18] and 5 mm for the latter [32]. The diameter of the bast fibre bundles is smaller than that of softwood, but the tensile strength is three times greater. The kenaf bast fibre has the potential as a reinforcing fibre in thermoplastic composites because of its superior toughness and high aspect ratio in comparison with other fibres.

2.4 Types of Natural Fibres

Fig. 2.15 Long kenaf fibres

Fig. 2.16 Random short kenaf fibres

Fig. 2.17 Kenaf filler

Fig. 2.18 Woven kenaf fibres

Fig. 2.19 A typical pulverizer to produce fillers (Courtesy of Innovative Pultrusion Sdn. Bhd, Seremban, Negeri Semblan, Malaysia)

According to Karnani et al. [17], a single fibre of kenaf can have a tensile strength and modulus as high as 11.9 and 60.0 GPa respectively. These properties can vary depending on the source, age and separating techniques of the fibre. Figures 2.15, 2.16, 2.17 and 2.18 show long kenaf fibres, random short kenaf fibres, kenaf filler, and woven kenaf fibres respectively. Kenaf is a light weight material, and its density is approximately 0.15 g/m^3. This material can easily be crushed to form fillers. Figure 2.19 shows a typical pulverizer to produce fillers. Its cellulose and lignin contents are approximately 32 and 25 %, respectively. Kenaf core has higher hemi-cellulose content than wood.

2.4 Types of Natural Fibres

Fig. 2.20 Beg from kenaf fibre (Courtesy of National kenaf and Tobacco Board, Malaysia)

In Malaysia, National kenaf and Tobacco Board (NKTB), Kubang Kerian, Kota Bharu, Kelantan established in April 2010 under the National kenaf and Tobacco Board Act 2009 [34]. It is now the policy of the government to replace tobacco with kenaf. Kenaf can be used as high quality bedding, woven and non-woven textiles, animal feed, oil absorption, fibre composite boards and paper [36] and bag (see Fig. 2.20).

2.4.4 Oil Palm Fibres

Oil palm (*Elaeis guineensis*) (see Fig. 2.21) is reported to originate from tropical forests in West Africa and it was introduced in Malaysia in 1870 [4]. Oil palm empty fruit bunch (EFB) (Fig. 2.22), oil palm frond (OPF), oil palm trunk (OPT), kernel shell, pressed fruit fibre (fruit mesocarp) and palm oil mill effluent (POME) generated from oil palm industry are regarded as waste and unutilized. These products normally caused major environmental pollution [14]. Lignocellulosic fibres can be extracted from OPT, OPF, fruit mesocarp and EFB [37]. Oil palm fibre (OPF) is extracted from EFB by retting process. The available retting processes are mechanical retting (hammering), chemical retting (boiling with chemicals), steam/vapour/dew retting and water/microbial retting [37]. Water retting is the most popular process among all those processes [30].

Fig. 2.21 Oil palm tree

Fig. 2.22 Empty fruit bunch (EFB) after palm oil extraction process

Oil palm fibre is hard and tough. However the presence of hydroxyl group made the fibres hydrophilic, leading to poor interfacial bonding with hydrophobic polymer in composites. This in turn, results in poor physical and mechanical properties of the composites [30].

Fig. 2.23 Sugar palm tree

2.4.5 Sugar Palm Fibres

Sugar palm (*Arenga pinnata*) is native to Indo-Malayan archipelago, it can be found in tropical South East and South Asia, Guam, Papua New Guinea [25]. Sugar palm (*Arenga pinnata*) is called by different names such as *kabung* or *enau* in Malaysia, *aren* in Indonesia and *gumoti* in India. Sugar palm fibre is a kind of natural fibre (in textile form) that comes from *Arenga pinnata* plant; a forest plant that can be found enormously in Southeast Asia like Indonesia and Malaysia. This fibre seems to have properties like other natural fibres, but the detail properties are not generally known yet. In Malaysia, sugar palm trees can be found in Bruas and Parit in Perak, Raub in Pahang, Jasin in Melaka and Kuala Pilah in Negeri Sembilan [27]. There are approximately 809 ha of sugar palm plantation found in Tawau, Sabah and 50 ha found in Benta, Pahang [33]. Figure 2.23 shows a sugar palm tree.

Generally, sugar palm fibre called *ijuk* in Malaysia has desirable properties like strength and stiffness and its traditional applications include paint brush, septic tank base filter, clear water filter, door mat, carpet, rope, cushion, roof material, broom, chair/sofa cushion, and for fish nest to hatch its eggs [39]. In certain regions in Indonesia, traditional application of sugar palm fibre includes handcraft for *kupiah* (Acehnese typical headgear used in prayer) and roofing for traditional house in Mandailing, North Sumatra, Indonesia. The sugar derived from the sugar palm tree is called palm sugar and it is one of the local delicacies widely consumed by Asians for making cakes, desserts, food coatings or mixed with drinks. It is

Fig. 2.24 Bali lamp from sugar palm fibre

Fig. 2.25 Sources of sugar palm fibres

produced by heating the sap derived from the sugar palm tree [13]. Sugar palm is proven to be acid and salt water resistant that made it feasible for use as rope used fisherman and domestic septic tank base filter. Figure 2.24 shows the use of sugar palm fibres in lamp in Bali, Indonesia. In fact, not only *ijuk* (natural woven cloth fibre) can be a useful source of fibres; other parts in sugar palm tree like sugar palm front (SPF), sugar palm bunch (SPB) and sugar palm trunk (SPT) can be used as fibres as shown in Fig. 2.25.

Fig. 2.26 Sugarcane trees

2.4.6 Sugarcane Fibres

Sugarcane (*Saccharum Officinarum*) is one of the major crops in Tropical region (Fig. 2.26). Total plantation area of sugar cane in Malaysia is nearly 34,500 acres [22]. Total plantation area of sugar cane in Malaysia is nearly 34,500 acres [22]. Sugarcane stalk (Fig. 2.27), from which bagasse fibres are derived, consists of an outer rind and an inner pith. Bagasse fibres are obtained after the extraction of the sugar-bearing juice from sugarcane [41]. Extracting sugar cane fibres from the plant stalks was considered to be a difficult and costly task [9]. Figure 2.28 shows a decortication machine to extract sugarcane juice from sugarcane juice maker in Malaysia. Residue of this sugarcane milling process gathered is a good source of sugarcane fibres.

Bagasse or sugar cane pulp fibres (sometimes called sugarcane bagasse) should be alkalinised, dried (Fig. 2.29), and milled before they can be used as high quality fibres [23]. Chiparus [9] reported that fibres in bagasse consist mainly of cellulose, pentosans, and lignin while [40] reported that chemical contents of bagasse are cellulose (35–40 %), natural rubber (20–30 %), lignin (15–20 %) and sucrose (10–15 %). Typically, the tensile strength of bagasse fibres is 70.85 MPa [8].

Utilization of sugarcane bagasse may contribute to environmental and economic development. Effort has been made to commercialize sugar cane fibres as useful products. Malaysia Airlines declared in 2012 that the meal box served on board from short haul flights to domestic and regional destinations are made from 100 % recyclable sugar cane fibre [1]. Bagasse has been used as a combustible

Fig. 2.27 Sugarcane stalk-source of bagasse fibres

Fig. 2.28 Sugarcane decortication machine

material for energy supply in sugar cane factories as in thermal power station in Guadeloupe (the French West Indies). Bagasse was also reported to be used in pulp and paper industries and for board materials [11]. Bagasse ash form bagasse fibres can be used as secondary filler in silica or carbon black filled natural rubber compound as reported by Kanking et al. [16].

2.4 Types of Natural Fibres

Fig. 2.29 Drying of bagasse fibres

Fig. 2.30 Pineapple tree

2.4.7 Pineapple Leaf Fibres

The scientific name of pineapple plant is *Ananas comosus L.* Pineapple is a long-leaf desert plant (Fig. 2.30) that can be grown in dry condition belonging to the Bromelicea family. The plant is normally grown in nurseries for the first year or so

Fig. 2.31 Fibre extraction by scrapping (Courtesy of Dr. Januar Parlaungan Siregar, Universiti Malaysia Pahang)

Fig. 2.32 Separating fibres from soft covering material (Courtesy of Dr. Januar Parlaungan Siregar, Universiti Malaysia Pahang)

and matures about 12–20 months old. The width of each leaf is about 50–75 mm. The fibres are contained in the spiky leaf of plant. Pineapple is a fibrous plant and it was reported that its fibres was as reinforcement or filler in composites. The majority of the research work carried out on pineapple leaf fibre (PALF) composites has been done in India and some South East Asian countries like Malaysia and Thailand. This could be due to the fact that the raw materials can be obtained there very cheaply, and so there is a great potential to commercialize this product and to enhance the quality of life of the people living in rural areas [2].

Conventional methods for PALF extraction include scraping (Fig. 2.31), retting and decorticating with a decorticator start from long fresh leaf and use mechanical force to remove soft covering material to provide long fibres (Figs. 2.32 and 2.33). In general, it is observed that, these methods produced low yield of coarse fiber bundles and up-scaling process is not easy to perform. Kengkhetkit and Amornsakchai [19] have come up with a new extraction method called mechanical milling. Apart from being used as reinforcement for composites, PALF was used as sound and thermal insulations. In Indonesia, PALF was used as raw material in textile industry (Fig. 2.34).

2.4 Types of Natural Fibres

Fig. 2.33 White and silky luster of PALF fibres (Courtesy of Dr. Januar Parlaungan Siregar, Universiti Malaysia Pahang)

Fig. 2.34 Shirt from pineapple leaf fibre

References

1. Anon.: Did you know? Malaysia Airlines Going Places, September, p. 14 (2012)
2. Arib, M.N., Sapuan, S.M., Hamdan, M.A.M.M., Paridah, M.T., Zaman, H.M.D.K.: Literature review of pineapple fibre reinforced polymer composites. Polym. Polym. Compos. **12**(4), 341–348 (2004)
3. Ashori, A.: Development of High Quality Printing Paper using Kenaf (Hibiscus Cannabinus) Fibres, Ph.D Thesis, Universiti Putra Malaysia (2004)
4. Bakar, A.A., Hassan, A.: Oil palm empty fruit bunch fibre-filled poly (vinyl chloride) composites. In: Salit, M.S. (ed.) Research on Natural Fibre Reinforced Polymer Composites, pp. 13–35. UPM Press, Serdang (2009)
5. Benítez, A.N., Monzón, M.D., Angulo, I., Ortega, Z., Hernández, P.M., Marrero, M.D.: Treatment of banana fiber for use in the reinforcement of polymeric matrices. Measurement **46**, 1065–1073 (2013)
6. Bilba, K., Arsene, M.A., Ouensanga, A.: Study of banana and coconut fibers: botanical composition, thermal degradation and textural observations. Bioresour. Technol. **98**, 58–68 (2007)
7. Biswas, S., Srikanth, G., Nangia, S.: Development of Natural Fibre Composites in India. www.tifac.org.in/news/cfq.htm (2006)
8. Cao, Y., Goda, K., Shibata, S.: Development and mechanical properties of bagasse fiber reinforced composites. Adv. Compos. Mater **16**, 283–298 (2007)
9. Chiparus, O.I.: Bagasse Fiber For Production Of Nonwoven Materials, Ph.D Dissertation, Louisiana State University (2004)
10. Georgopoulos, S.T., Tarantili, P.A., Avgerinos, E., Andreopoulos, A.G., Koukios, E.G.: Thermoplastic polymers reinforced with fibrous agricultural residues. Polym. Degrad. Stab. **90**, 303–312 (2005)
11. Ghazali, M.J.: Characterisation of natural fibres (sugarcane bagasse) in cement Composites. In: Proceedings of the World Congress on Engineering 2008 Vol II (WCE 2008), 2–4 July, London (2008)
12. Guimarães, J.L., Frollini, E., da Silva, C.G., Wypych, F., Satyanarayana, K.G.: Characterization of banana, sugarcane bagasse and sponge gourd fibers of Brazil. Ind. Crops Prod. **30**, 407–415 (2009)
13. Ho, C.W., Aida, W.M.W., Maskat, M.Y., Osman, H.: Changes in volatile compounds of palm sap (*Arenga pinnata*) during the heating process for production of palm sugar. Food Chem. **102**, 1156–1162 (2007)
14. Husin, M., Zakaria, Z.Z., Hassan, A.H.: Potentials of oil palm by-products as raw materials for agro-based industries. In: Proceedings of the National Symposium on Oil Palm By-products for Agrobased Industries, Kuala Lumpur, pp. 7–15 (1985)
15. Joseph, K., Filho, R.D.T., James, B., Thomas, S., De Carvalho, L.H.: A review on sisal fiber reinforced polymer composites. Revista Brasileira de Engenharia Agricola e Ambiental **3**, 367–379 (1999)
16. Kanking, S., Niltui, P., Wimolmala, E., Sombatsompap, N.: Use of bagasse fiber ash as secondary filler in silica or carbon black filled natural rubber compound. Mater. Des. **41**, 74–82 (2012)
17. Karnani, R., Krishnan, M., Narayan, R.: Biofiber-reinforced polypropylene composites. Polym. Eng. Sci. **37**, 476–483 (1997)
18. Kawai, S.: Summary Note of the 20th Meeting of Wood Adhesion Working Group, Akita (1999)
19. Kengkhetkit, N., Amornsakchai, T.: Utilisation of pineapple leaf waste for plastic reinforcement: 1. A novel extraction method for short pineapple leaf fiber. Ind. Crops Prod. **40**, 55–61 (2012)

References

20. Lai, C.Y.: Mechanical Properties and Dielectric Constant of Coconut Coir-Filled Propylene. Master of Science Thesis, Universiti Putra Malaysia, Serdang, Selangor, Malaysia (2004)
21. Lai, C.Y., Sapuan, S.M., Ahmad, M., Yahya, N., Dahlan, K.Z.H.M.: Mechanical and electrical properties of coconut coir fibre-reinforced polypropylene composites. Polym. Plast. Technol. Eng. **44**, 619–632 (2005)
22. Lee, S.C., Mariatti, M.: The effect of bagasse fibers obtained (from rind and pith component) on the properties of unsaturated polyester composites. Mater. Lett. **62**, 2253–2256 (2008)
23. Leite, J.L., Pires, A.T.N., Ulson de Souza, S.M.A.G., Ulson de Souza, A.A.: Characterisation of a phenolic resin and sugar cane pulp composite. Braz. J. Chem. Eng. **21**, 253–260 (2004)
24. Luo, S., Netravali, N.: Mechanical and thermal properties of environmental-friendly "green" composites made from pineapple leaf fibres and poly (hydroxybutyrate-valerate) resin. Polym. Compos. **20**, 367–378 (1999)
25. Mogea, J., Seibert, B., Smits, W.: Multipurpose palms: the sugar palm (*Arenga pinnata* (Wurmb) Merr.). Agrofor. Syst. **13**, 111–119 (1991)
26. Oksman, K., Skrifvars, M., Selin, J.F.: Natural fibres as reinforcement in polylactic acid (PLA) composites. Compos. Sci. Technol. **63**, 1317–1324 (2003)
27. Othman, A.R., Haron, N.H.: Potensi industri kecil tanaman enau. In: Nik, A.R. (ed.) Forest Research Institute of Malaysia (FRIM) Report. FRIM Press, Kepong, Malaysia (1992)
28. Paridah, M.T., Shukur, N.A.A., Harun, J., Abdan, K.: Kenaf- a journey towards energizing the biocomposite industry in Malaysia. In: Paridah, M.T., Abdullah, L.C., Kamaruddin, N. (eds.) Kenaf: Biocomposites, Derivatives and Economics, pp. 1–28. Pustaka Prinsip Sdn. Bhd, Kuala Lumpur (2009)
29. Paul, N.G.: Some methods for the utilization of waste from fiber crops and fiber wastes from other crops. Agric. Wastes **2**, 313–318 (1980)
30. Raju, G., Ratnam, C.T., Ibrahim, N.A., Rahman, M.Z.A., Yunus, W.M.Z.W.: Enhancement of PVC/ENR blend properties by poly(methyl acrylate) grafted oil palm empty fruit bunch fiber. J. Appl. Polym. Sci. **110**, 368–375 (2008)
31. Reddy, N., Yang, Y.: Biofibers from agricultural byproducts for industrial applications. Trends Biotechnol. **23**, 22–27 (2005)
32. Rymsza, T.A.: Utilization of kenaf raw materials. http:// www.hempology.org (1999)
33. Sahari, J., Sapuan, S.M., Zainudin, E.S., Maleque M.A.: Sugar palm tree: a versatile plant and novel source for biofibres, biomatrices, and biocomposites. Polym. Renew. Resour. **3**, 61–78 (2012)
34. Salleh, I.M.: Penanaman, penghasilan dan pengkomersilan kenaf: cabaran dan halatuju, Presented at LRGS Workshop on Kenaf: Sustainable Materials in Automotive Industry, 25–28 Dec. Tok Bali, Kelantan, Malaysia (2012)
35. Sapuan, S.M., El-Shekeil, Y.A.: Properties of kenaf fiber-reinforced elastomer composites. In: Proceedings of the Third International Conference of Institution of Engineering and Technology Brunei Darussalam (IETBIC 2012), Bandar Sri Begawan, 17–18 Sept, p. 25 (2012)
36. Seoung, T.K.: Fibre Reinforced Plastic Composite: Kenaf (Hibiscus cannabinus L.) Fibre-Polypropylene Blend, M.S. Thesis, Universiti Putra Malaysia, Serdang, Selangor, Malaysia (2002)
37. Shinoj, S., Visvanathan, R., Panigrahi, S., Kochubabu, M.: Oil palm fiber (OPF) and its composites: a review. Ind. Crops Prod. **33**, 7–22 (2011)
38. Stevens, E.S.: Green Plastics, an Introduction to the New Science of Biodegradable Plastics. Princeton University Press, Princeton (2002)
39. Suwartapraja, O.S.: Arenga pinnata: A case study of indigenous knowledge on the utilization of a wild food plant in West Java. www.geocities.com/inrik/opan.html (2003)
40. Vilay, V., Mariatti, M., Taib, R., Todo, M.: Effect of fiber surface treatment and fiber loading on the properties of bagasse fiber–reinforced unsaturated polyester composites. Compos. Sci. Technol. **68**, 633–638 (2008)

41. Wirawan, R.: Thermo-Mechanical Properties of Sugarcane Bagasse-Filled Poly(Vinyl Chloride) Composites, Ph.D Thesis, Universiti Putra Malaysia, Serdang, Selangor, Malaysia (2011)
42. Zainudin, E.S.: Effect of Banana Pseudostem Filler and Acrylic Impact Modifier on Thermo-Mechanical Properties of Unplastisized Poly(vinyl Chloride) Composites, Ph.D Thesis, Universiti Putra Malaysia, Serdang, Selangor, Malaysia (2009)

Chapter 3
Biopolymer

Abstract In this chapter a study of biopolymer is presented. Biopolymer is classified into five categories in this book namely polylactide, thermoplastic starch, cellulose, polyhydroxyalkanoates and synthetic biopolymer.

Keywords Biopolymer · Starch · Polylactide · Synthetic biopolymer · Cellulose

3.1 Introduction

Petroleum based polymers are extremely stable and commonly used in various industries include food packaging, construction, electric and electronic, furniture and automotive. One of the rapidly growing areas for the plastic usage is in packaging industry. Convenience, lightweight, safety, low price and aesthetically pleasing are the most important factors in determining the rapid growth in the manufacture of plastic packing. However, the wastes that come from petroleum based polymer material has brought negative impact not only for human being, but also unfortunately create the serious environmental problems. In addition, the shortage and high cost of fossil resources require alternative resources that are sustainable for our future generation. To address this challenge, it is therefore necessary to find out another raw material that can used for packaging industry and biodegradable polymers that come from sustainable source are now being considered as an alternative to the existing petrochemical-based polymers [16, 33]. Biopolymers are polymers that are biodegradable. The input materials for the production of these polymers may be either renewable (based on agricultural plant or animal products) or synthetic as shown in Fig. 3.1.

This chapter is written together with J. Sahari and Mohd Kamal Mohd Shah of Universiti Malaysia Sabah, Malaysia.

© Springer Science+Business Media Singapore 2014
M.S. Salit, *Tropical Natural Fibre Composites*,
Engineering Materials, DOI 10.1007/978-981-287-155-8_3

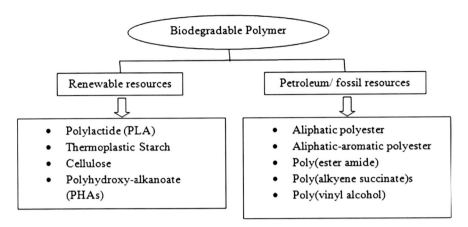

Fig. 3.1 Classifications of biopolymer

3.2 Classification of Biopolymers

3.2.1 Polylactide (PLA)

Polylactide is a thermoplastic biopolymer derived from renewable resources such as corn starch, tapioca and sugarcane. Its basic monomer is lactic acid, which is derived from starch by fermentation and it is a sustainable alternative to petrochemical-derived products, since the lactides produced from agricultural by-products mainly the carbohydrate rich substances [20, 21]. Lactic acid is a naturally occurring organic acid that can be produced by chemical synthesis or fermentation. Lactic acid was polymerised to poly(lactic acid), either by gradual polycondensation or by ring opening polymerisation [12, 17]. However, the most common way to obtain high-molecular-weight poly lactic acid is through ring opening polymerization of lactide. This process was carried out most commonly by using stannous octoate catalyst derived from stannous octoate and alcohol, was proposed as the substance initiating the polymerization through coordinative insertion of lactide [14]. Lactic acid can be synthesized into high-molecular-weight PLA by two different routes of polymerization [18, 30]. Lactic acid is condensation polymerized to yield a low-molecular-weight, brittle, glassy polymer, which, for the most part, is unusable for any applications unless external coupling agents are used to increase the molecular weight of the polymer. The molecular weight of this condensation polymer is low due to the viscous polymer melt, the presence of water and impurities. The second route of producing PLA is to collect, purify, and ring-open polymerize lactide to yield high-weightaverage molecular weight (M_w. 100,000) PLA. The high-molecular-weight PLA was done by a process where lactic acid and catalyst are azeotropically dehydrated in a refluxing, high-boiling, aprotic solvent under reduced pressures to obtain PLA with weight-average molecular weights >300,000 [9, 19, 24, 37]. This mechanism does

3.2 Classification of Biopolymers   41

not generate additional water, and hence, a wide range of molecular weights is accessible. Actually, there are three stereoforms of lactide are possible: L-lactide, D-lactide, and meso-lactide.

Recently, PLA meets many requirements as packaging applications and when plasticized with its own monomers, PLA becomes increasingly flexible, high-strength and high-modulus. So, a continuous series of products can be prepared by using PLA that can mimic PVC, LDPE, LLDPE, PP, and PS. Table 3.1 shows the starchy and cellulosic materials used for the production of lactic acid.

3.2.2 Thermoplastic Starch (TPS)

Nowadays, investigations are focussed to the development and characterization of biodegradable films from natural polymers since conventional synthetic plastic materials are resistant to microbial attack and biodegradation [2, 11]. Among all natural biopolymers, starch has been considered as one of the most promising one because of its easy availability, biodegradability, lower cost and renewability [49]. Starch is the major form of stored carbohydrate in plants such as corn, wheat, rice, and potatoes which composed of a mixture of two polymers of α-glucose i.e. linear amylose and a highly branched amylopectin. Amylose molecules consist of 200–20,000 glucose units which form a helix as a result of the bond angles between the glucose units. Amylopectin is a highly branched polymer containing short side chains of 30 glucose units attached to every 20–30 glucose units along the chain. Amylopectin molecules may contain up to two million glucose units as shown in Fig. 3.2 [39].

Starch is a natural polymer which occurs as granules in plant tissue, from which it can easily be recovered in large quantities. It is obtained from potatoes, maize, wheat and tapioca and similar sources. Starch can be modified in such a way that it can be melted and deformed thermoplastically. The resulting material is thus suitable for conventional plastic forming processes such as injection moulding and extruding. Starches from various sources are chemically similar and their granules are heterogeneous with respect to their size, shape, and molecular constituents. Proportion of the polysaccharides amylose and amylopectin become the most critical criteria that determine starch behaviour [8, 52]. Most amylose molecules (molecular weight $\sim 10^5$–10^6 Da) are consisted of $(1 \rightarrow 4)$ linked α-D-gluco-pyranosyl units and formed in linear chain. But, few molecules are branched to some extent by $(1 \rightarrow 6)$ α-linkages [5, 48]. Amylose molecules can vary in their molecular weight distribution and in their degree of polymerization (DP) which will affect to their solution viscosity during processing, and their retrogradation/recrystallization behavior, which is important for product performance. Meanwhile, amylopectin is the highly branched polysaccharide component of starch that consists of hundreds of short chains formed of α-D-glucopyranosyl residues with $(1 \rightarrow 4)$ linkages. These are interlinked by $(1 \rightarrow 6)$-α-linkages, from 5 to 6 % of which occur at the branch points [1]. As a result, the amylopectin shows the high

Table 3.1 Starchy and cellulosic materials used for the production of lactic acid

Substrate	Microorganism	Lactic acid yield
Wheat and rice bran	Lactobacillus sp.	129 g/l
Corn cob	Rhizopus sp. MK-96–1196	90 g/l
Pretreated wood	Lactobacillus delbrueckii	48–62 g/l
Cellulose	Lactobacillus coryniformis ssp. torquens	0.89 g/g
Barley	Lactobacillus caseiNRRLB-441	0.87–0.98 g/g
Cassava bagasse	L. delbrueckii NCIM 2025, L casei	0.9–0.98 g/g
Wheat starch	Lactococcus lactis ssp. lactis ATCC 19435	0.77–1 g/g
Whole wheat	Lactococcus lactis and Lactobacillus delbrueckii	0.93–0.95 g/g
Potato starch	Rhizopus oryzae, R. arrhizuso	0.87–0.97 g/g
Corn, rice, wheat starches	Lactobacillus amylovorous ATCC 33620	<0.70 g/g
Corn starch	L. amylovorous NRRL B-4542	0.935 g/g

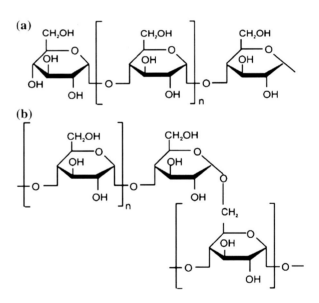

Fig. 3.2 Unit structure of **a** amylose and **b** amylopectin

molecular weight (10^7–10^9 Da) and its intrinsic viscosity is very low (120–190 ml/g) because of its extensively branched molecular structure.

Usually, a small amount of starch (6–30 %) was used as filler to increase the biodegradability of reinforced synthetic polymer [4, 10]. The starch is not a real thermoplastic, but, it will be act as synthetic plastic in the presence of a plasticizer (water, glycerol, sorbitol, etc.) at high condition of temperature. The various properties of thermoplastic starch product such as mechanical strength, water

3.2 Classification of Biopolymers

solubility and water absorption can be prepared by altering the moisture/plasticizer content, amylose/amylopectin ratio of raw material and the temperature and pressure in the extruder [32].

Starch is attractive because it is a cheap material and has very fast biodegradation rate. Under high temperature and shear, starch can be processed into a moldable thermoplastic, known as thermoplastic starch (TPS). Thermoplastic starch is plasticized starch that has been processed (typically using heat and pressure) to completely destroy the crystalline structure of starch to form an amorphous thermoplastic starch as shown in Fig. 3.3. During the gelatinization process, water contained in starch and the added plasticizers play an indispensable role because the plasticizers can form hydrogen bonds with the starch, replacing the strong interactions between the hydroxyl groups of the starch molecules, and thus making starch thermoplastic. In most literature for thermoplastic starch, polyols as plasticizers usually were used such as glycerol, sorbitol and urea.

Plasticizers are the most important material to increase the flexibility and processibility of TPS. There are large number of researches that have been performed on the plasticization of TPS using glycerol [41], sorbitol [15], urea, formamide [31], dimethyl sulfoxide [34] and low molecular weight sugars [22]. The properties of TPS also depend a lot on moisture. As water has a plasticizing power, the materials behavior changes according to the relative humidity of the air through a sorption-desorption mechanism [46]. At the same time, the properties of the materials evolve as time goes by, even moisture and temperature are controlled, translating into a lower elongation and higher rigidness.

Starch will act as biopolymer in the presence of a plasticizer such as water, glycerol and sorbitol at high temperature. Previously, the characterization of sugar palm starch as a biopolymer has been done by using the glycerol as plasticizers [41]. From the research, it was found that the tensile strength of SPS/G30 showed the highest value 2.42 MPa compared to the other concentration of the plasticizer. The higher the concentration of the plasticizers, the higher the tensile strength of plasticized SPS and optimum concentration was 30 wt%. The tensile strength decrease to 0.5 MPa when concentrations of plasticizer was 40 wt%. As the plasticizer content increased to 40 %, not enough SPS to be well bonded with glycerol and thus poor adhesion occurred which reduce the mechanical properties of SPS/G40.

Generally, as the plasticizer increase, the tensile strength and elongation of plasticized SPS increase, while the tensile modulus decreased. This phenomanaidicates that the plasticized SPS is more flexible when subjected to tension or mechanical stress. It is also most likely affect the crystallinity of starch by decreasing the polymer interaction and cohesiveness. Thus, this make the plasticized SPS become more flexible with the increasing of glycerol as shown in Figs. 3.4 and 3.5.

For the thermal properties, the Tg of dry SPS reaches 242.14 °C and decreased with addition of glycerol. The sample with high glycerol concentrations showed lower Tg values and Tg of starch without plasticizer were higher than those of samples with glycerol (Table 3.2). In the absence of plasticizers, starch are brittle. The addition of plasticizers overcomes starch brittleness and improves its flexibility and extensibility of the polymers.

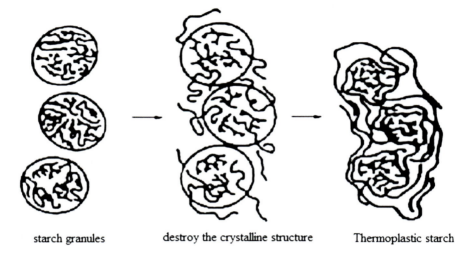

Fig. 3.3 Starch gelatinization process

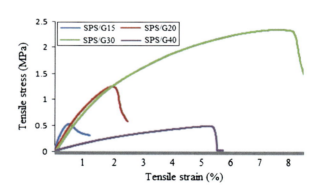

Fig. 3.4 Tensile stress-strain of plasticized SPS [41]

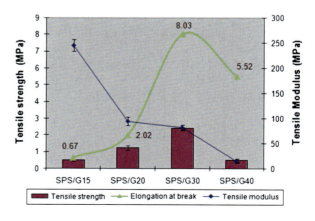

Fig. 3.5 Tensile properties of plasticized SPS [41]

3.2 Classification of Biopolymers

Table 3.2 Glass transition temperature (Tg) of plasticized SPS

Sample	Glass transition temperature, Tg (onset) (°C)	Glass transition temperature, Tg (midpoint) (°C)
Native SPS	237.91	242.14
SPS/G15	225.68	229.26
SPS/G20	206.44	217.90
SPS/G30	189.57	187.65
SPS/G40	176.71	177.03

3.2.3 Cellulose

The production of biomass is around 172 billion tons/year of which lignocellulosics in forests represent 82 % [7] and almost tropical plant is widely distributed in Southeast Asia [23, 42, 50]. Cellulose is the major constituents that are composed in lignocellulosics in spite of hemicelluloses, lignin, and extractives [44]. Hence, cellulose is the most abundant natural homopolymer and considered to be one of the most promising renewable resources and an environmentally friendly alternative to products that derived from the petrochemical industry. The cellulose content of such vegetable matter varies from plant to plant. For example, oven-dried cotton contains about 90 % cellulose, while an average wood has about 50 %. The balance is composed of lignin, polysaccharides other than cellulose and minor amounts of resins, proteins and mineral matter. Plant derived cellulose has been widely used as either reinforcement or matrix. Many cellulose derivatives have been prepared of which the esters and ethers are important as shown in Fig. 3.6.

The cellulose esters are useful polymers for the manufacture of plastics and the most important of esters is cellulose acetate. This material has been extensively used in the manufacture of films, moulding and extrusion compounds, fibres and lacquers. The methods available to produce acetate may be considered under two headings, homogeneous acetylation, in which the acetylated cellulose dissolves into a solvent as it is formed, and the heterogeneous technique, in which the fibre structure is retained. For the preparation of the acetate homogeneous acetylation, it is considered in three stages which are pretreatment of the cellulose, acetylation and hydrolysis. The detail explanation has been discussed. Nowadays, a wide range of cellulose acetate compounds are commercially available and the properties of these compounds are depended on three major factors:

(1) The chain length of the cellulose molecule.
(2) The degree of acetylation.
(3) The type and amount of plasticiser(s).

The greater the molecular weight the higher is the flow temperature and the heat distortion temperature. However, the variations in molecular weight have less

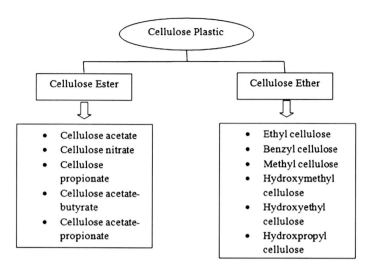

Fig. 3.6 Classification of cellulose derivatives

effect than the variations in the degree of acetylation and in the plasticiser used. It is observed that increasing the degree of acetylation will obviously reduce the hydroxyl content and this will increase the water resistance. In addition, it is that proved that an increase in the degree of acetylation reduces the hardness, impact strength and water absorption.

According to Offerman [40] cellulose ester based structures can be "triggered" to biodegrade and some grades of cellulose esters are readily biodegradable to environment (e.g. gases, water, and biomass) [26]. Many cellulose ester derevatives such as acetates, propionates and butyrates can be burned without generating toxic products or residues, thus allowing their use in the incineration of waste for recovery of heat value. However, not all cellulose derivatives degrade at appreciably fast rates to be considered "biodegradable." Toriz et al. [44] has been claim that the rate of biodegradation increases as long as the degree of substitution and the length of the side chain substituent decreases. Generally, different type of fillers, plasticizers, and other additives that are used can also influence biodegradation rate of cellulose ester.

In spite of cellulose acetate, the reaction between cellulose and nitric acid is also one of esterification and this reaction will produce cellulose nitrate. It is possible to achieve varying degrees of esterification according to the number of hydroxyl groups that have been replaced by the nitrate group. This material is not made commercially but esters with lower degrees of nitrate are most important and typically used in plastics, lacquers films and lacquers cordite industry.

3.2.4 Polyhydroxyalkanoates (PHAs)

Polyhydroxyalkanoates are biodegradable polymers produced by prokaryotic organisms from renewable resources. This biological polyesters are mainly produced by microbial fermentation processes [35]. The starting material for Polyhydroxibutyrate is made from sucrose or starch by a process of bacterial fermentation. Varying the nutrient composition of the bacteria produces differences in the end product. This makes it possible to tune the properties of the material, e.g. its moisture resistance. The polymer can be formed by injection, extrusion, blowing and vacuum forming. The production of PHAs by submerged fermentation processes has been intensively studied over the last 30 years and recently, alternative strategies have been proposed, such as the use of solid-state fermentation or the production of PHAs in transgenic plants. Due to the large impact of the carbon source price on production costs, one of the most important approaches to reduce costs is to use wastes and by-products as raw material for the fermentation process. The producing of PHAs from waste and by-products materials has the potential to reduce PHA production costs, because the raw material costs contribute a significant part of production costs in traditional PHA production processes [28].

The first PHA to be identified was poly(3-hydroxybutyrate) [P(3HB)] from Bacillus megaterium Lemoigne [29] and Wallen and Rohwedder [47] identified PHAs from sewage sludge with different monomer units : 3-hydroxyvalerate (3HV) and 3-hydroxyhexanoate (3HHx). It has been reported that, approximately 150 different hydroxyalkanoic acids are known to be incorporated into polyhydroxyalkanoates [43] and over 90 genera microbial species have being reported to accumulate these polyesters [51]. Among that species, *Cupriavidus necator* previously known as *Ralstonia eutropha*, *Hydrogenomonas eutropha*, *Alcaligenes eutrophus*, *R. eutropha* and *Wautersia eutropha* is the most studied and most frequently used.

PHAs can be classified as short-chain-length or medium-chain-length polymers depending on the length of their monomers. Kim and Lenz [25] have been reported that the short-chain-length group includes polyesters containing monomers that are C4 or C5 hydroxyalkanoic acids, such as poly(3-hydroxybutyrate) [P(3HB)], poly(3-hydroxyvalerate) [P(3HV)] or the copolymer P(3HB-co-3HV). The second group which is medium-chain-length includes those polymers formed by monomers equal to or longer than C6. These group usually existing as copolymers of two to six different types of 3-hydroxyalkanoic acid units. Beside that, Fukui et al. [13] have been done the hybrid polymers containing both short-chain and medium-chain monomer units such as poly(3-hydroxybutyrate-co-3-hydroxyhexanoate) synthesized by *Aeromonas caviae*. The properties of PHAs vary considerably and the chain length of the monomer play a vital aspect that influences polymer hydrophobicity, melting point, glass transition temperature and degree of crystallinity [6]. Table 3.3 shows the properties of different PHAs.

48 3 Biopolymer

Table 3.3 The properties of different PHAs

Polymer	Melting point (°C)	Young modulus (GPa)	Tensile strength (MPa)	Elongation at break (%)
P(3HB)	175–180	3.5–4	40	3–8
P(3HB-co-3HV) (3 mol % HV)	170	2.9	38	–
P(3HB-co-3HV) (20 mol % HV)	145	1.2	32	50–100
P(3HB-co-4HV) (3 mol % HV)	166	–	28	45
P(3HB-co-4HV) (10 mol % HV)	159	–	24	242
P(3HO)	61	–	6–10	300–450

3.2.5 Synthetic Biopolymer

The relatively high price of biodegradable polymers of synthetic substances, e.g. aliphatic aromatic copolyesters has prevented them from reaching a large scale market. The best known application is for making substrate mats. Synthetic compounds derived from petroleum can also be a starting point for biodegradable polymers, e.g. aliphatic aromatic copolyesters. These polymers have technical properties resembling those of polyethylene (LDPE). Although these polymers are produced from synthetic starting materials, they are fully biodegradable and compositable.

Besides being available on a sustainable basis, biopolymers have several economic and environmental advantages. Biopolymers could also prove an asset to waste processing. For example, replacing the polyethylene used in coated papers by a biopolymer could help eliminate plastic scraps occurring in compost. Consumers have a lively interest in biopolymers too. Conventional plastics are often seen as environmentally unfriendly. Sustainable plastics could therefore provide an image advantage.

The major advantage of biodegradable packaging is that it can be composted. But the biodegradability of raw materials does not necessarily mean that the product or package made from them (e.g. coated paper) is itself compositable. Biopolymers can also have advantages for waste processing. Coated paper (with e.g. polyethylene) is a major problem product for composting. Although such materials are usually banned from inclusion in organic waste under separate collection schemes, some of them usually end up nonetheless in the mix. The paper decomposes but small scraps of plastic are left over in the compost. The adoption of biopolymers for this purpose would solve the problem.

Biodegradability is not only a function of origin but also of chemical structure and degrading environment. Sometimes thermoset bioresins, even if made or derived from bioresources, may not be biodegradable. PLA as well as PHAs are renewable resource-based biopolyesters, in contrast to PCL, PBS and aliphatic-aromatic polyesters, which are petroleum-based biodegradable polyesters. Aliphatic polyesters are readily biodegradable, whereas aromatic polyesters like poly(ethylene terephthalate) (PET), are nonbiodegradable. However, aliphatic-aromatic copolyesters have been shown to be biodegradable, and recently these polyesters have gained commercial interest, especially for packaging applications [45]. Eastman's Easter Bio® and BASF's Ecoflex® are two examples of aliphatic-aromatic copolyesters based on butanediol, adipic acid, and terephthalic acid. Eastar Bio is highly linear in structure while Ecoflex has a long-chain branched structure.

As early as 1973, the biodegradability of PCL was demonstrated [27]. It is a tough and semirigid material at room temperature with a modulus between low-density and high-density polyethylene. 3 PCL has a low melting point ($\sim$60 °C), low viscosity, and can be melt processed easily. It possesses good water, oil, solvent, and chlorine resistance. PCL is widely used as a blending partner with a number of polymers, especially with hydrophilic starch plastic [3, 39]. Biocomposites from PCL and natural fibers have been developed. The tensile strength and Young's modulus of PCL improved by 450 and 115 %, respectively, after reinforcement with 40 wt% wood flour. Poly(alkylene dicarboxylate) biodegradable aliphatic polyesters have been developed by Showa Highpolymer under the trade name Bionolle®. Different grades of Bionolle include polybutylene succinate (PBS), poly(butylene succinate-cobutylene adipate) (PBSA), and poly(ethylene succinate). SK Chemicals also produces aliphatic PBS polyesters. DuPont's biodegradable Biomax copolyester resin, a modified form of PET, was launched in 1997. Its properties, according to DuPont, are diverse and customizable, but they are generally formulated to mimic polyethylene or polypropylene [36].

3.3 Conclusions

Biodegradable polymers which are poly lactide (PLA), thermoplastic starch (TPS), cellulose and polyhydroxyalkanoates, will certainly play an important role in the plastics industries in the future due to their biodegradability and come from renewable resources as well. Their offers a possible alternative to the traditional non-biodegradable polymers especially due to recycling process is difficult and not economical. Current and future developments in biodegradable polymers and renewable input materials focus relate mainly to the scaling-up of production and improvement of product properties. Larger scale production will increase availability and reduce prices. Currently either renewable or synthetic starting materials may be used to produce biodegradable polymers. Two main strategies may be followed in synthesizing a polymer. One is to build up the polymer structure from a monomer by a process of chemical polymerization. The alternative is to take a naturally occurring polymer and chemically modify it to give it the desired properties. A disadvantage of chemical modification is however that the biodegradability of the polymer may be adversely affected. Therefore it is often necessary to seek a compromise between the desired material properties and biodegradability.

Finally, biodegradable polymer to meet wide range of applications especially in food packaging, fast-food restaurants, electric and electronic devices, furniture and automotive industries. So, further research must be done so that mechanical and other properties can be easily manipulated depending on the end-users requirements.

References

1. Anonymous: Making the most of starch. http://www.rsc.org/Education/EiC/issues/2006Sept/MakingMostStarch.asp (2006). Accessed 26 Apr 2011
2. Arvanitoyannis, I.: Totally-and-partially biodegradable polymer blends based on natural and synthetic macromolecules: preparation and physical properties and potential as food packaging materials. J. Macromol. Sci. **39**(2), 205–271 (1999)
3. Averous, L., Moro, L., Dole, P., Fringant, C.: Properties of thermoplastic blends: starch-polycaprolactone. Polymer **41**, 4157–4167 (2000)
4. Bagheri, R.: Effect of processing on the melt degradation of starch-filled polypropylene. Polym. Int. **48**, 1257–1263 (1999)
5. Buleon, A.: Starch granules: structure and biosynthesis. Int. J. Biol. Macromol. **23**(2), 85–112 (1998)
6. Crank, M., Patel, M., Marscheider-Weidemann, F., Schleich, J., Hüsing, B., Angerer, G.: Techno-economic feasibility of large-scale production of bio-based polymers in Europe (PRO-BIP). Final Report Prepared for the European Commission's Institute for Prospective Technological Studies (IPTS) (2004)
7. Hon, D.N.-S.: Embrapa instrumentação agropecuária. In: Frollini, E., Leão, A.L., Mattoso (eds.) Natural Polymers and Agrofibers Composites, p. 292. USP-IQSC, São Carlos (2000)
8. Ellis, R.P.: Starch production and industrial use. J. Sci. Food Agric. **77**(3), 289–311 (1998)
9. Enomoto, K., Ajioka, M., Yamaguchi, A.: U.S. Pat. No. 5,310,865 (1994)
10. Evangelista, R.L., Nikolov, Z.L., Sung, W., Jane, J., Gelina, R.J.: Effect of compounding and starch modification on properties of starch-filled low density polyethylene. Ind. Eng. Chem. Res. **30**, 1841–1846 (1991)
11. Fang, J., Fawler, P., Eserig, C., González, R., Costa, J., Chamudis, L.: Development of biodegradable laminate films derived from naturally occurring carbohydrate polymers. Carbohydr. Polym. **60**(1), 39–42 (2005)
12. Farrington, D.W., Davies, J.L., Blackburn, R.S.: Poly(lactic acid) fibers. In: Blackburn, R.S. (ed.) Biodegradable and Sustainable Fibers, pp. 191–220. Woodhead Publishing, Cambridge (2005)
13. Fukui, T., Shiomi, N., Doi, Y.: Expression and characterization of (*R*)-specific enoyl coenzyme A hydratase involved in polyhydroxyalkanoate biosynthesis by *Aeromonas caviae*. J. Bacteriol. **180**, 667–673 (1998)
14. Garlotta, D.: A literature review of poly(lactic acid). J. Polym. Environ. **9**, 63–84 (2001)
15. Gaudin, S., Lourdin, D., Le Botlan, D., Ilari, J.L., Colonna, P.: Plasticization and mobility in starch-sorbitol film. J. Cereal Sci. **29**, 273–284 (1999)
16. Gerngross, T.U., Slater, S.C.: Biopolymers and the environment. Science **299**(3), 822–825 (2003)
17. Gupta, B., Revagade, N., Hilborn poly(lactic acid) fiber: an overview. Prog. Polym. Sci. **32**, 455–482 (2007)
18. Hartmann, M.H.: In: Kaplan, D.L. (ed.) Biopolymers from Renewable Resources, pp. 367–411. Springer, Berlin (1998)
19. Ichikawa, F., Kobayashi, M.,. Ohta, M, Yoshida, Y., Obuchi, S., Itoh, H.: U.S. Pat No. 5,440,008 (1995)

References

20. John, R.P., Gangadharan, D., Nampoothiri, K.M.: Genome shuffling of *Lactobacillus delbrueckii* mutant and *Bacillus amyloliquefaciens* through protoplasmic fusion for L-lactic acid production from starchy wastes. Bioresour. Technol. **99**, 8008–8015 (2008)
21. John, R.P., Nampoothiri, K.M., Pandey, A.: Solid-state fermentation for L-lactic acid production from agro wastes using *Lactobacillus delbrueckii*. Process Biochem. **41**, 759–763 (2006)
22. Kalichevsky, M.T., Jaroszkiewicz, E.M., Blanshard, J.M.V.: A study of the glass transition of amylopectin-sugar mixtures. Polymer **34**, 346–358 (1993)
23. Kampeerapappun, P., Phattararittigul, T., Jittrong, S., Kullachod, D.: Effect of chitosan and mordants on dyeability of cotton fabrics with *Ruellia tuberosa Linn*. Chiang Mai J. Sci. **38**, 95–104 (2010)
24. Kashima, T., Kameoka, T., Higuchi, C., Ajioka, M., Yamaguchi, A.: U.S. Pat No. 5,428,126 (1995)
25. Kim, Y.B., Lenz, R.W.: Polyesters from microorganisms. In: Babel, W., Steinbüchel, A. (eds.) Biopolyesters, pp. 51–79. Springer, Berlin (2001)
26. Komarek, R.J., Gardner, R.M., Buchanan, C.M., Gedon, S.C.: Aerobic biodegradation of cellulose acetate. J. Appl. Polym. Sci. **50**, 1739 (1993)
27. Leaversuch, R.: Biodegaradable polyester: packaging goes green. Polym. Plast. Technol. Eng. **48**, 66 (2002)
28. Leda, R., Castilho, D., Mitchell, A., Denise, M.G.: Freire. Production of polyhydroxyalkanoates (PHAs) from waste materials and by-products by submerged and solid-state fermentation. Bioresour. Technol. **100**, 5996–6009 (2009)
29. Lemoigne, M.: Produits de dehydration et de polymerisation de l'acide ß-oxobutyrique. Bull. Soc. Chim. Biol. **8**, 770–782 (1926)
30. Lunt, J.: Large-scale production, properties and commercial applications of polylactic acid *polymers*. Polym. Degrad. Stab. **59**, 145–152 (1998)
31. Ma, X., Yu, J., Jin, F.: Urea and formamide as a mixed plasticizer for thermoplastic starch. Polym. Int. **53**, 1780–1785 (2004)
32. Mohanty, A.K., Misra, M., Hinrichsen, G.: Biofibres, biodegradable polymers and biocomposites: an overview. Macromol. Mater. Eng. **276**(277), 1–24 (2000)
33. Mohanty, A.K., Misra, M., Drzal, L.T., Selke, S.E., Harte, B.R., Hinrichsen, G.: Natural fibres, biopolymers, and biocomposites: an introduction. In: Natural Fibres, Biopolymers and Biocomposites, pp. 1–36. CRC Press, Boca Raton (2005)
34. Nakamura, S., Tobolsky, A.V.: Viscoelastic properties of plasticized amylose films. J. Appl. Polym. Sci. **11**, 1371–1381 (1967)
35. Nath, A., Dixit, M., Bandiya, A., Chavda, S., Desai, A.J.: Enhanced PHB production and scale up studies using cheese whey in fed batch cultures of *Methylobacteria* sp. ZP24. Bioresour. Technol. **99**, 5749–5755 (2008)
36. Nitz, H., Semke, H., Landers, R., Mulhaupt, R.: Reactive extrusion of polycaprolactone compounds containing wood flour and lignin. J. Apply Polym. Sci. **81**, 1972–1984 (2001)
37. Ohta, M., Yoshida, Y., Obuchi, S.: U.S. Pat No. 5,444,143 (1995)
38. Potts, J.E., Clendinning, R.A., Ackart, W.B., Niegish, W.D.: Biodegradability of synthetic polymers. Polym. Sci. Technol. **3**, 61 (1973)
39. Ray, S.S., Bousmina, M.: Biodegradable polymers and their silicate nanocomposites: In greening the 21st century materials world. Prog. Mater Sci. **50**(8), 962–1079 (2005)
40. Offerman, R.J.: U.S. Pat. No. 6,462,120 (2002)
41. Sahari, J., Sapuan, S.M., Zainudin, E.S., Maleque, M.A.: Thermo-mechanical behaviors of thermopolymer starch derived from sugar palm tree (*Arenga pinnata*). Carbohydr. Polym. **92**, 1711–1716 (2013)
42. Sangwichien, C., Sumanatrakool, P., Patarapaiboolchai, O.: Effect of filler loading on curing characteristics and mechanical properties of thermoplastic vulcanizate. Chiang Mai J. Sci. **35**, 141–149 (2008)
43. Steinbüchel, A.: Perspectives for biotechnological production and utilization of biopolymer: metabolic engineering of polyhydroalkanoate biosyenthesis pathways as a successful example. Macromol. Biosci. **1**, 1–24 (2001)

44. Toriz, G., Gatenholm, P., Seiler, B.D., Tindall, D.: Cellulose fiber-reinforced cellulose esters: biocomposites for the future. In: Natural Fibres, Biopolymers and Biocomposites, pp. 617–639. CRC Press, Boca Raton (2005)
45. Van der Walle, G.A.M., de Koning, G.J.M., Weusthuis, R.A., Eggink, G.: Properties, modifications and applications of biopolyesters. Adv. Biochem. Eng. **71**, 263–291 (2001)
46. Vilpoux, O., Avérous, L.: Starch-based plastics in technology, use and potentialities of Latin American starchy tubers. In: Cereda, M.P., Vilpoux, O. (eds.) NGO Raízes and Cargill Foundation, pp. 521–553. NGO Raízes and Cargill Foundation, São Paolo (2004)
47. Wallen, L.L., Rohwedder, W.K.: Biopolymers of activated sludge. Environ. Sci. Technol. **8**, 576–579 (1974)
48. Whistler, R.L., Daniel, J.R.: Molecular structure of starch. In: Whistler, R.L., BeMiller, J.N., Paschall, E.F. (eds.) Starch, Chemistry and Technology, 2nd edn, pp. 312–388. Academic Press, Inc, Orlando (1984)
49. Xu, Y., Kim, K., Hanna, M., Nag, D.: Chitosan–starch composite film: preparation and characterization. Ind. Crops Prod. **21**(2), 185–192 (2005)
50. Ya'acob, A.M., Sapuan, S.M., Ahmad, M., Dahlan, K.Z.M.: The mechanical properties of polypropylene / glass fiber composites prepared using different samples preparation methods. Chiang Mai J. Sci. **31**, 233–241 (2004)
51. Zinn, M., Witholt, B., Egli, T.: Occurrence, synthesis and medical application of bacterial polyhydroxyalkanoate. Adv. Drug Deliv. Rev. **53**, 5–21 (2001)
52. Zobel, H.F.: Molecules to granules: a comprehensive starch review. Starch **40**(2), 44–50 (1988)

Chapter 4
Mechanical and Other Related Properties of Tropical Natural Fibre Composites

Abstract In this chapter, studies on mechanical properties of selected tropical natural fibre reinforced polymer composites are presented. Banana, coconut, kenaf, oil palm, sugar palm, sugarcane and pineapple fibre reinforced polymer composites were the tropical natural fibre composites elaborated in this chapter. Tensile, flexural and impact properties were among the mechanical properties being studied. The effects of various parameters on mechanical properties of tropical natural fibre composites were investigated. These parameters include fibre contents, type of treatment agents, and fibre sizes. A section on fibre-matrix interfacial bonding including fibre treatments is also presented. Brief discussion on water absorption of natural fibre composites is also made available for the readers.

Keywords Mechanical properties · Tensile properties · Flexural properties · Impact properties · Fibre-matrix bonding

4.1 Introduction

Mechanical properties, that are the most commonly being studied, when considering tropical natural fibre composites, include strength, stiffness, fracture toughness and impact. Other properties include shear, compressive, fatigue and creep. Mechanical properties are the major properties being determined in many products particularly the components that are applied by certain amount of loads.

Tensile testing is the most common testing to determine tensile properties of natural fibre composites. Tensile properties, such as tensile strength and tensile modulus of natural fibre composites can be determined by tension tests using standards such as ASTM D3039. Preparation of specimens and tools are the initial steps in tensile testing. Universal testing machine (UTM) is normally used to carry out tensile testing (see Fig. 4.1). The composite specimen to be tested is placed at the top gripper of the UTM vertically. The cross-head speed for testing is entered

© Springer Science+Business Media Singapore 2014
M.S. Salit, *Tropical Natural Fibre Composites*,
Engineering Materials, DOI 10.1007/978-981-287-155-8_4

Fig. 4.1 Universal testing machine

and the load is applied to the specimen. The data and graph that show the properties of the specimen are printed out when the specimen breaks.

Flexural testing is used to determine the properties of a material as a response to combination of tensile and compressive stresses [42]. Flexural properties such as flexural strength and modulus can be determined by ASTM test method D790 using UTM at room temperature. Natural fibre composite specimen is prepared with rectangular cross section in either three-point bending mode or four-point bending mode.

Impact strength is a measure of fracture toughness of material subjected to impact load. It measures the ability of a material to absorb energy before fracture [8]. There are many impact testing methods available such as Izod and Charpy impact tests. The main contributors of impact properties in fibre composites are 1. The interfacial bonding strength, 2. The matrix properties, and 3. Fibre properties [51].

4.2 Banana Fibre Composites

Banana fibres can be obtained from pseudo-stem and leave after the crop harvest. Banana fibres are still classified as leaf fibre of vegetable fibre, even though they are extracted from pseudo-stem [4]. According to Inai [20] banana fibres have

Fig. 4.2 Comparison of experimental and theoretical tensile strengths of banana fibre reinforced phenol formaldehydes composites [21]

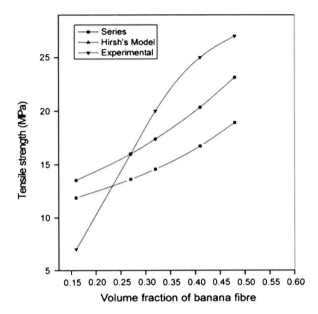

natural white colour with bright luster and are finer than sisal when properly extracted and dried.

Sapuan et al. [40] carried out tensile and flexural (three-point bending) investigation of woven banana pseudostem fibre reinforced epoxy composites. Three specimens of woven banana pseudostem fibre composites were made available for different geometries. Maximum stresses in x and y-directions were 14.14 and 3.398 MN/m^2 respectively. Young's moduli in x and y-directions were 0.976 and 0.863 GN/m^2 respectively. The maximum flexural stress and flexural modulus in x-direction were 26.181 MN/m^2 and 2.685 GN/m^2, respectively.

Joseph et al. [22] carried out experimental work to determine mechanical properties of banana fibre reinforced phenol formaldehyde composites and compared the results with those of glass fibre reinforced phenol formaldehyde composites. Figure 4.2 shows the results of comparison of tensile strengths of banana fibre reinforced phenol formaldehyde composites theoretically and experimentally. From this figure, it can be learnt that the tensile strength of banana fibre composites increased with the increase of fibre loading.

Zainudin et al. [56] studied the effect of addition of banana pseudo-stem fibres by weight in unplasticized poly(vinyl chloride) (UPVC) composites on mechanical properties of the composites. The tensile properties studied include thermal, flexural and impact properties. The specimens were prepared using compression moulding process. Undesirable results were obtained in that the higher the fibre contents, the lower the values of tensile strength. Neat UPVC) specimens gave the highest tensile strength. It is because of the presence of weak region where the fibres and matrix was interfaced. Hydrophilic banana pseudo-stem fibres and hydrophobic UPVC were not compatible. Similar trend was observed for flexural strength. For impact strength, as the fibre content is increased beyond neat polymer, the impact strength was also increased. However, after 30 % fibre loading, the impact strength showed decreasing trend. It was reported that at 30 % and above, fibre wetting was the major issue.

Adding the second fibre to the first fibre in polymer composites to bring the hybridization effect improved mechanical properties of the composites. Venkateshwaran et al. [50] reported that sisal fibre was added to banana fibre epoxy composites and improvement in terms of tensile strength and modulus, flexural strength and modulus and impact strength was observed. The higher the percentage of sisal, the better is the mechanical properties of the composites.

Inai [20] studied mechanical properties of hybrid banana pseudo-stem/glass fibre reinforced polyester composites. Composite specimens were prepared using compression moulding process at different fibre loadings. The increase in banana fibre loadings in the composites caused the decrease in the tensile and flexural properties of hybrid composites. The effect of hybridization was observed with the addition of glass fibre in the banana fibre composites in terms of the significant increase in the tensile, flexural and impact strengths of all composites.

Boopalan et al. [6] also carried similar study with hybridization of banana and jute fibres to reinforce epoxy composites and its effects on mechanical properties (tensile strength, flexural strength and impact strength) were investigated. Equal percentage of banana and jute fibres in composites (which made up of 30 % of fibres in composites) gave the maximum values of mechanical properties. Beyond 30 % of fibre content, the mechanical properties had reduced mainly due to poor interfacial adhesion between fibres and matrix.

4.3 Coconut Fibre Composites

Lai [25] studied mechanical properties of coconut fibre reinforced polypropylene composites. Fibre content, fibre size and different chemical treatment of coir fibres are three independent variables in the determination of mechanical properties coir fibre composites and composite specimens were prepared using compression moulding process. Composites with treated fibres gave better tensile and flexural moduli compared to untreated fibre composites.

Sapuan et al. [38] investigated tensile and flexural properties of coconut shell filled epoxy composites at three different filler loading, i.e. 5, 10 and 15 %. Composite specimens were prepared using hand lay-up process. It is observed as the filler content was increased in the composites, the values of tensile and flexural properties were also increased.

Sapuan et al. [39] investigated tensile and flexural strengths of coconut spathe-fibre reinforced epoxy composites. Composite specimens were fabricated by the hand lay-up process and the composites were prepared at 30 % fibre loading. The maximum tensile and flexural strengths for the coconut spathe fibre reinforced

4.3 Coconut Fibre Composites

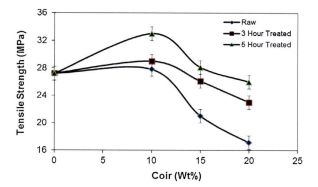

Fig. 4.3 Tensile strength of untreated and chemically treated coir fibre composites [31]

epoxy composites were 11.6 and 67.2 MPa respectively. The tensile and flexural strengths of coconut spathe fibre reinforced epoxy composites were much lower than other more established natural fibre composites. Fibre treatment can improve the interfacial bonding between fibres and matrix thus improving the strength properties.

The work of Mir et al. [31] was concerned with determination of mechanical properties of coir fibre reinforced polypropylene (PP) composites. Coir fibres were used either in untreated form or being chemically treated with chromium sulfate and sodium bicarbonate salt in acidic media and they were bonded with PP composites. It was concluded from this study that chemically treated coir fibre reinforced PP composites gave better mechanical properties than untreated coir fibre PP composites (see Fig. 4.3).

Lignin was used as additive in coir fibre reinforced PP composites and its effect on mechanical properties (tensile strength) was studied by Morandim-Giannetti et al. [29]. Lignin was added in the composites with and without compatibilizer (maleic anhydride grafted polypropylene, PP-g-MA). With the addition PP-g-MA in the composites along with lignin, the tensile strength of the composites was reduced, but without compatibilizer, lignin alone cannot change the tensile strength of the composites and it remained unaffected by it.

4.4 Kenaf Fibre Composites

Sapuan et al. [41] studied the effect of the period of soil burial on mechanical properties of kenaf fibre reinforced thermoplastic polyurethane (TPU) composites. Kenaf bast fibre reinforced TPU composites were prepared via melt-mixing method using internal mixer. Soil burial test was performed to study the effect of moisture absorption on mechanical properties of the composites. Tensile and flexural properties of the composites were determined before and after the soil burial tests for the maximum of 80 days. Tensile strength of kenaf fibre reinforced TPU composite reduced to 16.14 MPa after 80 days of soil burial test. It was also

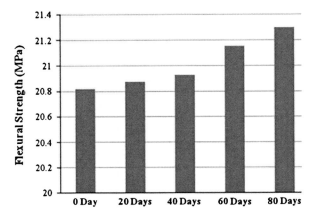

Fig. 4.4 Flexural strength of kenaf TPU composites after soil burial tests [41]

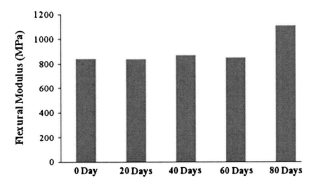

Fig. 4.5 Flexural modulus of kenaf TPU composites after soil burial tests [41]

observed that there was slight increase in flexural strength of soil buried kenaf fibre reinforced TPU composite specimens (see Fig. 4.4). For flexural modulus, the results have not changed except for the case of 80 days of soil burial (Fig. 4.5).

El-Shekeil et al. [11] and El-Shekeil [12] investigated the effect of fiber contents (20, 30, 40, and 50 % weight percent) on tensile, flexural, and impact properties of kenaf bast fiber reinforced TPU composites. The composite specimens were prepared using melt-mixing and compression moulding methods. The highest tensile strength was obtained 30 % fiber loading and the tensile modulus was increased with the increase in fibre content as shown in Fig. 4.6. There was improvement in flexural strength and modulus the increase in fibre loading as shown in Fig. 4.7. The increase in fibre loading caused the decrease in impact strength.

Russo et al. [38] reported their work on flexural and impact properties of alkali and silane treated kenaf fibre reinforced low density polyethylene (LDPE) composites. Composite samples were prepared using injection moulding process. Improvement in flexural and impact properties were observed for treated fibre composites compared to untreated fibre composites.

4.4 Kenaf Fibre Composites

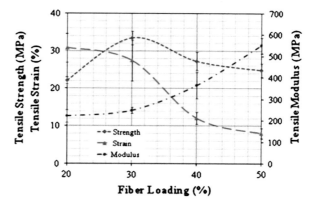

Fig. 4.6 Tensile properties of kenaf TPU composites [12]

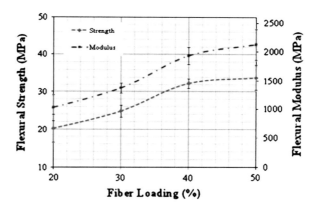

Fig. 4.7 Flexural properties of kenaf TPU composites [12]

Rezali [36] studied mechanical properties of untreated and alkali treated kenaf fibre reinforced epoxy composites and he compared the properties with ramie composites. Hand lay-up process was used to fabricate the composite specimens. Mechanical properties studied include tensile, flexural and impact properties. The impact properties study was quite unique because the study performed high velocity impact testing using Ballistic Automated Network Gun (BANG) at four different projectile speeds. For treated specimens, the fibres were treated with 6 % of sodium hydroxide (NaOH). The results of the study revealed that treated fibre ramie and kenaf composites gave better performance than untreated fibre composites in terms of impact properties by projectile travelling at low velocities. At low velocity, ramie composites gave better impact properties than kenaf composites but at high velocity, the results were just in the opposite.

Hanan [17] studied the effect of accelerated weathering on mechanical properties of kenaf fibre reinforced high density polyethylene (HDPE) composites. Mechanical properties of kenaf composites were measured after accelerated weathering tests were conducted for 1,000 h of exposure. Two types of composites were studied, i.e. unbleached kenaf fibre reinforced HDPE composite and its bleached counterpart. Ii was found that unbleached kenaf HDPE composites have better tensile strength than bleached HDPE composites because in bleached fibre, cellulose was more exposed and thus they had the tendency to absorb more moisture. As far as flexural strength is concerned, unbleached kenaf HDPE composites also show higher value and this was due to the higher lignin content in the composites. It provided better rigidity of the composites.

Rashdi [35] carried out extensive research work on moisture absorption capacity of kenaf fibre reinforced unsaturated polyester composites and to study its effect on mechanical properties of the composites. Kenaf fibre reinforced unsaturated polyester composites were subjected to different environmental tests i.e. water immersion, soil burial and natural weather tests. There was a decrease in tensile strength and modulus after the specimens were subjected to those three environmental tests for 4 months. Relative humidity of composite specimens after undergoing natural weather test was lower compared to the cases of water immersion and soil burial tests. This led to higher strength of the composites after undergoing natural weather test than those of soil burial and water immersion tests. This suggests that kenaf unsaturated polyester composites is suitable for outdoor applications.

4.5 Oil Palm Fibre Composites

Rozman et al. [37] investigated tensile performance of oil palm empty fruit bunch (EFB) fibre reinforced polyurethane composites. The fibres were used untreated or being treated with different types of isocyanates: hexamethylene diisocyanate (HMDI) and toluene diisocyanate (TDI). Treated fibre composites show higher tensile (Figs. 4.8 and 4.9) and flexural properties than untreated fibre composites. From the figures it is observed that tensile strength and modulus increased with the increase of fibre loading.

Yusoff [59] performed a study on mechanical properties of oil palm fibre reinforced polymer composites. Two polymer materials were studied; namely phenol formaldehyde (PF) and epoxy. Composite specimens were prepared using

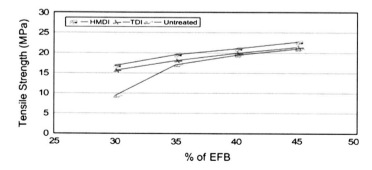

Fig. 4.8 The effect of the EFB loading and isocyanate treatment on the tensile strength [37]

4.5 Oil Palm Fibre Composites

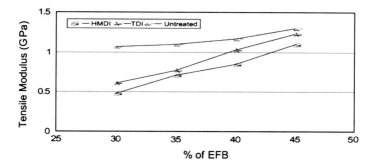

Fig. 4.9 The effect of the EFB loading and isocyanate treatment on the tensile modulus [37]

hand lay-up process. Oil palm EFB fibres were randomly chopped to form a reinforcement system. The results obtained are, for oil palm EFB fibre/PF composites, as the fibre loading was increased, the tensile and flexural strengths were also increased the highest values were found at 60 % fibre loading. But for EFB fibre/epoxy composites, the tensile and flexural strengths for all composites were lower than neat epoxy. The highest strengths for all fibre loadings were at 5 % fibre loading. The weak interfacial adhesion between fibres and matrix is believed to cause this problem.

Ibrahim [18] investigated the mechanical properties of oil palm ash filled unsaturated polyester composites. Composites with different filler loading were prepared (10, 20 and 30 % by weight) using hand lay-up process. Tensile and flexural strengths were decreased as the filler contents in the composites were increased. The decreasing trend of tensile strength might be due to the poor interfacial adhesion between oil palm ash filler and the matrix. However, tensile and flexural moduli demonstrated opposite results. Fu et al. [15] stated that the lower distribution of filler in polymer could decrease the tensile properties of the composites.

Jawaid et al. [24] reported that the hybridization of jute fibres with oil palm in epoxy composites decreased the impact strength of the composites. Khairiah and Khairul [27] carried out the study of mechanical properties of EFB fibre reinforced polyurethane (PU) composites. It was reported that the ratio of PU matrix to EFB fibres of 35:65 gave the highest values of hardness, and impact and flexural strengths. Abu Bakar et al. [2] reported that the EFB and oil palm front fibre reinforced epoxy composites demonstrated the values of impact strength as high as 40 % more than the neat epoxy.

4.6 Sugar Palm Fibre Composites

According to Ishak et al. [23] sugar palm fibre is unique for its durability and their resistance to sea water. In the past, ropes for ship cordages were made from sugar palm fibres because of their resistance against sea water. Nowadays, the use of sugar palm fibres is extended to other industries such as in building, and household appliances.

Sahari [40] studied mechanical properties of different morphological parts of sugar palm fibre reinforced polyester composites. Those sugar palm morphological parts include sugar palm frond (SPF), sugar palm bunch (SPB), sugar palm trunk (SPT) and sugar palm natural fabric (SPNF). SPNF is locally known in Malaysia as *ijuk*. For tensile strength, SPB reinforced polyester composites had the highest value. For flexural properties, SPT reinforced polyester composites were the highest. SPF reinforced polyester composites showed the highest value of impact strength.

Sahari et al. [39] studied the effect of fibre loading on mechanical properties of sugar palm fibre (SPF) reinforced plasticized sugar palm starch (SPF/SPS) composites. The biocomposites were prepared at 10, 20 and 30 % fibre loading by using glycerol as plasticizer for the starch. The mechanical properties (tensile and flexural properties) of plasticized starch improved with the increase of SPF (see Figs. 4.10 and 4.11). Scanning electron microscopy micrographs showed homogeneous distribution of fibres and matrix with excellent interfacial adhesion between them which in turn improved the mechanical properties of the composites.

Siregar [47] studied tensile and flexural properties of sugar palm fibre reinforced epoxy composites. Three types of fibre arrangements were selected: i. long random filaments where the filament came from the original form of sugar palm bundle, ii. chopped random filament, and iii. woven roving filament, where the weave filaments are in 0° and 90° directions. Hand lay-up process was employed to produce composite specimens. Fibre loadings used were 10, 15 and 20 %. The results showed that the increase in fibre loading had decreased the tensile and flexural strengths of composites reinforced with all three types of fibre arrangements. However, for tensile and flexural moduli results contrary to the strengths were obtained. Comparing three types of fibre arrangements, woven roving fibre

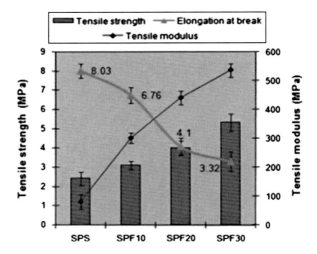

Fig. 4.10 Tensile properties of sugar palm fibre reinforced sugar palm starch composites [40]

4.6 Sugar Palm Fibre Composites

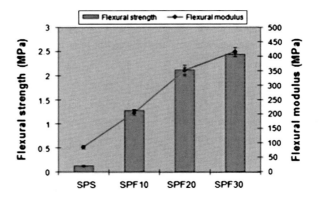

Fig. 4.11 Flexural properties of sugar palm fibre reinforced sugar palm starch composites [40]

composites showed the highest values of tensile and flexural strengths. Jusoh [26] also carried out research work similar to Siregar [47] using three different types of fibres (long random, chopped random and woven roving). She performed tensile testing of the composites and studied the morphology using Scanning Electron Microscopy of the fractured specimens of the composites and found that woven roving fibre composites demonstrated good interfacial adhesion among three different composites.

In the early work of Bachtiar [6], mechanical properties of alkali-treated sugar palm fibre reinforced epoxy composites were evaluated. Hand lay-up process was employed to produce composite specimens. In this study, the fibres were treated by alkali solution with 0.25 and 0.5 M NaOH solution for 1, 4 and 8 h of soaking time. Tensile, flexural and impact properties of composites improved with the alkali treatment. There were no consistent results in different mechanical properties when the soaking was increased.

In another work of Bachtiar [7], he investigated mechanical properties of short sugar palm fibre reinforced high impact polystyrene composites. Composite specimens were prepared using melt mixing and compression moulding method. Tensile, flexural and impact testing were performed at fibre content of 40 %. Alkali treatment and compatibilizing agent were employed to resolve compatibility issue. Alkali treatment using 4 and 6 % of sodium hydroxide (NaOH) and 2 and 3 % of compatibilizing agent; polystyrene-block-poly(ethylene-ran-butylene)-block-poly(styrene-graft-maleic-anhydride) were used. The results showed that NaOH solution and compatibilizing agent improved the tensile strength, flexural strength, flexural modulus and impact strength of treated composites over untreated fibre composites.

Leman [29] also carried out experimental investigation on mechanical properties of sugar palm fibre reinforced epoxy composites. In his studies, fibres were subjected to various treatments before they were used as reinforcements in composites. Those treatment agents used include sea water, fresh (pond) water and sewage water. The treatments had modified the surface properties of the sugar palm fibres thus resulted in a better adhesion quality compared to the untreated fibres. This in turn, improved the mechanical properties of the composites.

A Master of Science work of Ishak [21] was concerned with determination of mechanical properties of treated and untreated woven sugar palm fibre reinforced unsaturated polyester composites. Woven fibres were obtained in original form from the tree. Compression moulding was used in the study. Two major issues were tried to be resolved by Ishak [21]. The first was to study the effect of fibre content (by different layers of fibres) on the mechanical properties of the composites, and the second was to study the effect of sea water treatment on the fibre surface on mechanical properties of the composites. In the former, the higher the fibre content (more layers were placed), the better the tensile strength and modulus, flexural strength and impact strength. For the latter, the sea water treatment improvement the mechanical properties of the composites.

A Ph.D work of Ishak [22] was on tensile properties of impregnated composites made of sugar palm fibre and unsaturated polyester. Tensile strength of the composites was studied with the fibre loadings were varied from 10 to 50 %. Tensile strength of the composites improved with the increased of fibre loading from 10 to 30 % and above 30 %, the strength reduced. The decrease was associated with the poor fibre-matrix wettability due to intermingling of fibres in the composites. It resulted in insufficient resin to wet the fibres.

A recent work by Ibrahim [19] was on the effects of flame-retardant agents on mechanical properties of impregnated sugar palm fibre reinforced unsaturated polyester composites. Flame retardant agents used in the form of fillers were aluminium trihydroxide (ATH) and magnesium hydroxide (MH). The effects of flame retardant agents for fibre loading of 10–50 % on tensile strength and modulus of the composites were investigated. The tensile strength and modulus of composites filled with ATH were higher than those composites filled with MH. However, for both composites filled with ATH and MH, as the fibre loading was increased, the tensile and flexural strengths were also decreased.

4.7 Sugarcane Fibre Composites

Many investigations were carried out on the mechanical properties of bagasse fibre reinforced polymer composites in the past. The matrices used include both thermoplastic and thermosetting polymers [56] such as polypropylene, unsaturated polyester, polyethylene (PE), poly(ethylene vinyl acetate) (EVA) and poly(vinyl chloride) (PVC).

Mariatti and Abdul Khalil [31] performed research investigation on the properties of bagasse fibre reinforced unsaturated polyester composites. The composite specimens were prepared using vacuum bag moulding process. The treatment agents in this study were sodium hydroxide (NaOH) and acrylic acid (AA). The effect of fibre treatments on flexural strength and modulus of the composites was investigated. Fibre treatment using both NaOH and AA had improved the flexural properties of the composites.

4.7 Sugarcane Fibre Composites

Wirawan et al. [57] studied mechanical properties of composites of sugarcane bagasse in poly(vinyl chloride) (PVC) matrix produced by a compression moulding method. He determined mechanical properties of both rind (outer) and pith (inner) parts of bagasse fibre reinforced PVC composites and found that tensile strength and modulus of rind/PVC composites were higher than neat PVC but the impact energy was lower than neat polymer. Wirawan [56] also studied mechanical properties of rind bagasse fibre PVC composites with the fibres were treated with benzoic acid, sodium hydroxide (NaOH), and poly-[methylene(polyphenyl)isocyanate] (PMPPIC) as coupling agent. From these chemical treatments, composites with the fibres treated with PMPIC demonstrated the highest tensile strength and modulus. One interesting finding is that composites with unwashed (contained sugar), untreated bagasse had the highest tensile properties than chemically treated fibre composites. There is possibility that sugar contributed towards improving the strength and modulus of the composites.

Figures 4.12 and 4.13 show the tensile strengths and tensile moduli results of bagasse filled polyethylene (PE) composites based on the experiments reported by Agunsoye and Aigbodion [3]. Two types of fillers were studied i.e. uncarbonized

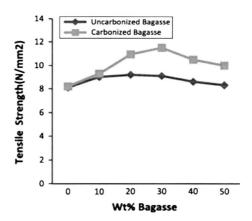

Fig. 4.12 Tensile strength of bagasse fibre reinforced PE composites [3]

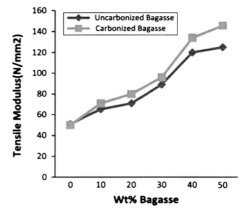

Fig. 4.13 Tensile modulus of bagasse fibre reinforced PE composites [3]

(UBp) and carbonized (CBp) bagasse particles and the bagasse particles were filled in the composites at filler loading of 10, 20, 30, 40 and 50 wt%. The tensile strengths of the composite increased with the increase of filler loading and the maximum strengths were obtained at 20 wt% for UBp composites and 30 wt% for CBp composites.

Vallejos et al. [53] carried out research on the mechanical properties (tensile properties) bagasse fibre reinforced thermoplastic starch (TPS) composites. Bagasse fibres were obtained from ethanol–water fractionation of bagasse. TPSs were derived from corn and cassava. The increase in bagasse fibres in the composites improved tensile strength and modulus of the composites. Cao et al. [9] studied tensile, flexural and impact strengths of non-treated and alkali treated bagasse fibre reinforced biodegradable aliphatic polyester composites. The tensile and impact strengths of the untreated bagasse fibre composites increased with increase in fibre content (up to 65 % fibre content). Lower percentage of alkali treatment agent (1 % of NaOH) in composites demonstrated the highest values of tensile, flexural and impact strengths, compared with higher percentages of alkali (3 and 5 % of NaOH). Stael et al. [50] performed experimental work on impact properties of bagasse composites and at different volume fraction and fibre size. The matrix system used were poly(ethylene vinyl acetate) poly(ethylene vinyl acetate) (EVA), polypropylene and polyethylene. The results showed that the incorporation of bagasse strongly improved the impact properties of EVA polymer. The impact strength was independent of the bagasse size, but varied with the volume fraction.

4.8 Pineapple Fibre Composites

PALFs have traditionally been used as threads and textile materials in the past. The use of PALF polymer composites in engineering field faced the major problem of high moisture absorption and poor interfacial adhesion between hydrophilic fibres and hydrophobic polymer. A number of studies were conducted in the recent years gearing towards the possible use of these materials in various industries.

Threepopnatkul et al. [51] studied the effect of surface treatment (sodium hydroxide) on mechanical properties of pineapple leaf fibre (PALF) reinforced polycarbonate composites. Surface treatment of PALF in composites demonstrated improved mechanical properties (tensile strength, Young's modulus and impact strength). Arib et al. [4] investigated the tensile and flexural properties of PALF–polypropylene composites at different fibre loadings. Figures 4.14 and 4.15 show the comparison of experimental and theoretical tensile strengths and moduli of PALF PP composites [4]. Experimentally, as the fibre content increased, the tensile strength was also increased but beyond certain percentages, it eventually dropped. The reason could be due to the poor interfacial bonding between fibres and matrix. Similar results were observed for tensile moduli.

4.8 Pineapple Fibre Composites

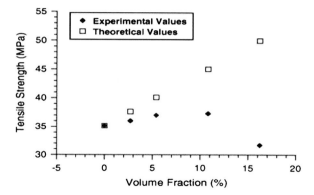

Fig. 4.14 Tensile strength versus fibre loading of PALF/PP composites [4]

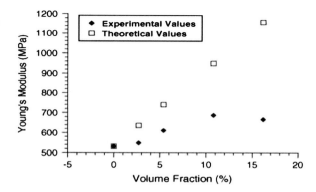

Fig. 4.15 Young's modulus versus fibre loading of PALF/PP composites [4]

Siregar [49] studied the effects of various treatments on mechanical properties of PALF reinforced high impact polystyrene (HIPS) composites. One of the treatment agents was electron beam irradiation. Two types of crosslinking agent or polyfunctional monomer were used for electron beam irradiation of PALF reinforced HIPS composites and they were trimethylolpropane triacrylate (TMPTA) and tripropylene glycol diacrylate (TPGDA). The composites were prepared using mixer and compression moulding. The composite sheets were irradiated using an electron beam machine at the dose range of 0–100 kGy. All specimens were irradiated at room temperature with accelerator energy of 3 MeV, beam current of 2 mA at 20 kGy/pass. The mechanical properties studied were tensile and flexural properties. Tensile strength and modulus of PALF reinforced HIPS composites had increased with addition of polyfunctional monomer compared to untreated fibre composites. The tensile strength increased with the increase of irradiation dose. This was true for both TMPTA and TPGDA. For flexural strength and modulus, both TMPTA and TPGDA addition increased the strength up to certain done. At higher dose, both flexural strength and modulus showed decreasing trend.

Sapuan and Siregar [44] and Siregar and Sapuan [48] investigated mechanical properties of pineapple leaf (PALF) fibre reinforced high impact polystyrene (HIPS) composites. In the former study, the effect of alkali treatments on mechanical properties of PALF reinforced HIPS composites was investigated. The fibre size selected was 10–40 meshes. The fibres were treated with sodium hydroxide (NaOH) at three different concentrations (0, 2 and 4 %). The untreated PALF fibr/HIPS composites had the lowest tensile strength and modulus. The higher the alkali concentration, the higher were the tensile strength and modulus. For flexural properties, the latter used compatibilizing agents to improve mechanical properties of the composites and they were poly-styrene-block-poly(ethylene-ran-butylene)-block-poly(styrene-graft-maleic anhydride) (PSgMA) and poly(styrene-co-maleic anhydride) (PScoMA). The PALF fibres were incorporated into the HIPS using a mixer and the composites were prepared in compression moulding machine. The addition of both compatibilizers had improved tensile strength and modulus, flexural strength and modulus, impact strength and hardness of PALF-HIPS composites. However, PScoMA performed better than PSgMA in improving the mechanical properties of the composites.

Mohamed [33] reported his work on mechanical properties of PALF reinforced vinyl ester composites where the manufacturing method used was compression moulding. The fibre was in the form of randomly oriented non-woven PALF mat. It was noted that using PALF in vinyl ester either in the large size (large vascular bundles) or smaller size (fine fibre bundle strands) did not affect flexural strength and modulus of the composites.

Liu et al. [30] studied the effect of fibre loading and compatibilizer on mechanical properties of PALF biopolymer composites where biopolymer was derived from soya. Composites were prepared using twin-screw extrusion and injection moulding processes. The compatibilizer used was polyester amide grafted glycidyl methacrylate (PEA-g-GMA). The mechanical properties (tensile properties, flexural properties and impact strength) of the composites increased with the increase in fibre content and with the presence of the compatibilizer. The increase in mechanical properties of the composites was attributed to the presence of the compatibilizer; due to the interactions between hydroxyl groups in the PALF and epoxy groups in PEA-g-GMA.

George et al. [16] reported that addition of PALF in low-density polyethylene caused the composites to demonstrate increased water uptake and decreased mechanical properties. These weaknesses were overcome by fibre treatments using NaOH, silane A-172, isocyanate (PMPPIC) and benzoyl peroxide. These treatments helped in improving fibre-matrix adhesion and thus mechanical properties. Uma Devi et al. [52] reported the work on mechanical properties of PALF reinforced polyester composites. The fibres were in the form of randomly oriented chopped and the composites were fabricated using hand lay-up process. It was stated that the optimum length of fibre was 30 mm. The study revealed that tensile strength and modulus, flexural strength and impact strength increased linearly with the increase in fibre loading.

4.9 Fibre-Matrix Interface in Tropical Natural Fibre Composites

The major problem associated with natural fibre and polymer matrix is low compatibility of hydrophilic natural fibres with hydrophobic polymer. This problem leads to low strength and poor fibre-matrix interfacial adhesion and poor mechanical properties in general. fibres have polar hydroxyl groups on the surface in the forms of cellulose and lignin and these are not compatible with non polar polymer matrix and therefore well bonded interface cannot be achieved. Natural fibre composites also have undesirable water absorption that is causing swelling and creating voids. Water absorption normally takes place at the free hydroxyl groups on the cellulose chains.

Many methods were developed to enhance the adhesion between fibres and matrix in natural fibre composites which could be in the category of physical methods or chemical treatment methods. For improving compatibility of between matrix and fibres the methods that have been reported to be used include surface cleaning, etching or ablation, cross linking or branching agent, modification of surface chemical structure and introduction of free radical, natural cold plasma, hybrid yarn, thermo-treatments, calendering, electric discharge method stretching, compatibilizing agents, alkali treatment (mercerization) with sodium hydroxide (Naoh), acetylation, silane treatment with silane coupling agent), permanganate treatment, stearic acid treatment, peroxide treatment, acrylation, dewaxing, bleaching, heat treatment, cranoethylation, isocyanate treatment, benzoylation, (fibre treatment with benzyl chloride), fibre grafting with maleic anhydride modified PP (MAPP) grafting with acrylonitrile and methyl methacrylate, surface modification, Scouring, chemical modification, and polymeric coating.

Explanation of some of the methods mentioned above is given here. Acetylation is a treatment that is common with cellulose to form a hydrophobic thermoplastic and has the potential to have the same results on natural fibers. It is also known as esterification method for plasticizing of cellulose fibres. Coupling agent is an important modification method that can improve interfacial adhesion of fibres and matrix and an example of it is silane coupling agent. Silane chemical coupling (graft copolymerization) is used as a coupling agent and it is a treatment commonly used in glass composite production and is starting to find uses in natural fiber composites. The silane molecules that have bifunctional groups react with fibre and polymer matrix to form a bridge between them thus effectively couple the fibre and matrix [58]. According to Abdelmouleh et al. [1] silane coupling agents offered three major advantages:

(i) they are commercially available in a large scale;
(ii) at one end, they bear alkoxysilane groups capable of reacting with OH-rich surface, and
(iii) at the second end, they have a large number of functional groups which can be tailored as a function of the matrix to be used.

Permanganate treatment is performed using potassium permanganate ($KMnO_4$) solution in acetone at different concentrations. The peroxide-induced grafting of polyethylene sticks to the surface of the fibre and the peroxide that started free radicals reacted with the hydroxyl group of the composite constituents. Heat treatments of natural fibres increase durability and dimensional stability of composites but reduce mechanical properties.

4.10 Water/Moisture Absorption

Water absorption in composites is the amount of water absorbed by composites as a function of time. Natural fibre composites normally absorb water due to the hydrophilic nature of natural fibres. Water uptake in natural fibre composites may cause mechanical properties to suffer after they are being exposed to water or moisture for extended periods of time. Water absorption study is carried out using water immersion test by directly immersing polymer and natural fibre polymer composites in distilled water at ambient temperature in accordance with ASTM D570.

The water absorption of natural fibre composites is normally calculated as follows:

$$WA = ((m_1 - m_0)/m_0) \times 100\%$$

where m_0 is the original weight of the natural fibre composites, and m_1 is the weight of the natural fibre composites after being immersed in the water. Generally the water absorption of the natural fibre composites increased with the increase in natural fibre contents in composites. Fickian diffusion is normally used to study the water absorption of natural fibre composites. When the water absorption curve fits the linear Fickian diffusion curve, the water absorption process is considered a Fickian diffusion. In a typical linear Fickian diffusion curve, the water absorption of natural fibre composites increased linearly as a function of time of water soaking during the early period of immersion; then it reached a plateau region and finally the curve became constant. Water absorption in natural fibre composites can be reduced by the use of coupling agent in the fibre matrix interface. It is distributed on the surface of fibre. The coupling agent blocked the water permeation into the fibre [14].

References

1. Abdelmouleh, M., Boufi, S., Belgacem, M.N., Dufresne, A.: Short natural-fibre reinforced polyethylene and natural rubber composites: Effect of silane coupling agents and fibres loading. Compos. Sci. Technol. **67**, 1627–1639 (2007)
2. Abu Bakar, M.A., Natarajan, V.D., Kalam, A., Nor Hayati, K.: Mechanical properties of oil palm fibre reinforced epoxy for building short span bridge. In: Proceedings of the 13th International Conference on Experimental Mechanics, Alexandroupolis, Greece, 1–6 July, pp. 97–98 (2007)

References

3. Agunsoye, J.O., Aigbodion, V.S.: Bagasse filled recycled polyethylene bio-composites: Morphological and mechanical properties study. Results Phys. **3**, 187–194 (2013)
4. Arib, R.M.N., Sapuan, S.M., Ahmad, M.M.H.M., Paridah, M.T., Khairul Zaman, H.M.D.: Mechanical properties of pineapple leaf fibre reinforced polypropylene composites. Mater. Des. **27**, 391–396 (2006)
5. ASTM International: ASTM D123-09, Standard Terminology Relating to Textiles. ASTM International, West Conshohocken (2009)
6. Bachtiar, D.: Mechanical Properties of Alkali-Treated Sugar Palm (Arenga Pinnata) Fibre Reinforced Epoxy Composites, Master of Science Thesis, Universiti Putra Malaysia, Serdang, Selangor, Malaysia (2008)
7. Bachtiar, D.: Mechanical and Thermal Properties of Short Sugar Palm (Arenga Pinnata Merr.) Fibre-Reinforced High Impact Polystyrene Composites, Ph.D Thesis, Universiti Putra Malaysia, Serdang, Selangor, Malaysia (2012)
8. Boopalan, M., Niranjanaa, M., Umapathy, M.J.: Study on the mechanical properties and thermal properties of jute and banana fiber reinforced epoxy hybrid composites. Compos. B Eng. **51**, 54–57 (2013)
9. Cao, Y., Shibata, S., Fukumoto, I.: Mechanical properties of biodegradable composites reinforced with bagasse fibre before and after alkali treatments. Compos. A Appl. Sci. Manuf. **37**, 423–429 (2006)
10. Callister, W.D.: Materials Science and Engineering: An Introduction, 7th edn. Wiley, New York (2007)
11. El-Shekeil, Y.A., Sapuan, S.M., Abdan, K., Zainudin, E.S.: Influence of fiber content on the mechanical and thermal properties of kenaf fiber reinforced thermoplastic polyurethane composites. Mater. Des. **40**, 299–303 (2012)
12. El-Shekeil, Y.A.: Preparation and Characterization of Kenaf Fibre Reinforced Thermoplastic Polyurethane Composites, Ph.D Thesis, Universiti Putra Malaysia, Serdang, Selangor, Malaysia (2012)
13. Fang, H., Zhang, Y., Deng, J., Rodrigue, D.: Effect of fiber treatment on the water absorption and mechanical properties of hemp fiber/polyethylene composites. J. Appl. Polym. Sci. **127**, 942–949 (2012)
14. Fu, S.Y., Feang, X.Q., Lauke, B., Mai, Y.W.: Effects of particle size, particle/matrix interface adhesion and particle loading on mechanical properties of particulate-polymer composites. Compos. B Eng. **39**, 933–961 (2008)
15. George, J., Bhagawan, S.S., Thomas, S.: Effect of environment on the properties of low-density polyethylene composites reinforced with pineapple fibres. Compos. Sci. Technol. **58**, 1471–1485 (1998)
16. Hanan, A.U.: Effect of Accelerated Weathering on Kenaf-Reinforced High Density Polyethylene Composite, Master of Science Thesis, Universiti Putra Malaysia, Serdang, Selangor, Malaysia (2012)
17. Ibrahim, M.S.: Physical and Thermomechanical Properties of Oil Palm Ash-Filled Unsaturated Polyester Composites, Master of Science Thesis, Universiti Putra Malaysia, Serdang, Selangor, Malaysia (2012)
18. Ibrahim, A.H.: Effects of Flame-Retardant Agents on Mechanical Properties and Flammability of Impregnated Sugar Palm Fibre-Reinforced Polymer Composites, Master of Science Thesis, Universiti Putra Malaysia, Serdang, Selangor, Malaysia (2013)
19. Inai, N.H.: Mechanical and Physical Properties of Hybrid Banana Pseudostem/Glass Fibre Reinforced Polyester Composites, Master of Science Thesis, Universiti Putra Malaysia, Serdang, Selangor, Malaysia (2013)
20. Ishak, M.R.: Mechanical Properties of Treated and Untreated Woven Sugar Palm Fibre-Reinforced Unsaturated Polyester Composites, Master of Science Thesis, Universiti Putra Malaysia, Serdang, Selangor, Malaysia (2009)

21. Ishak, M.R.: Enhancement of Physical Properties of Sugar Palm (Arenga Pinnata Merr.) Fibre-Reinforced Unsaturated Polyester Composites Via Vacuum Resin Impregnation, Ph.D Thesis, Universiti Putra Malaysia, Serdang, Selangor, Malaysia (2012)
22. Ishak, M.R., Sapuan, S.M., Leman, Z., Rahman, M.Z.A., Anwar, U.M.K. Siregar, J.P.: Sugar palm (Arenga pinnata): its fibres, polymers and composites. Carbohydr. Polym. **91**, 699–710 (2013)
23. Jawaid, M., Abdul Khalil, H.P.S., Abu Bakar, A., Hassan, A., Dungani, R.: Effect of jute fibre loading on the mechanical and thermal properties of oil palm–epoxy composites. J. Compos. Mater. **47**, 1633–1641 (2012)
24. Joseph, S., Sreekala, M.S., Oommen, Z., Koshy, P., Thomas, S.: A comparison of the mechanical properties of phenol formaldehyde composites reinforced with banana fibres and glass fibres. Compos. Sci. Technol. **62**, 1857–1868 (2002)
25. Jusoh, S.M.: A Case Study on Tensile Properties and Morphology of Arenga Pinnata Fiber Reinforced Epoxy Composites, Master of Science Thesis, Universiti Putra Malaysia, Serdang, Selangor, Malaysia (2006)
26. Khairiah, B., Khairul, A.M.A.: Biocomposites from oil palm resources. J. Oil Palm Res. (Special Issue), 103–113 (2006)
27. Lai, C.Y.: Mechanical Properties and Dielectric Constant of Coconut Coir-Filled Propylene, Master of Science Thesis, Universiti Putra Malaysia, Serdang, Selangor, Malaysia (2004)
28. Leman, Z.: Mechanical Properties of Sugar Palm Fibre-Reinforced Epoxy Composites, Ph.D Thesis, Universiti Putra Malaysia, Serdang, Selangor, Malaysia (2009)
29. Liu, W., Misra, M., Askeland, P., Drzal, L.T., Mohanty, A.K.: 'Green' composites from soy based plastic and pineapple leaf fiber: fabrication and properties evaluation. Polymer **46**, 2710–2721 (2005)
30. Mariatti, J., Abdul Khalil, H.P.S.: Properties of bagasse fibre-reinforced unsaturated polyester (USP) composites. In: Sapuan, S.M. (ed.) Research on Natural Fbre Reinforced Polymer Composites, pp. 63–83. UPM Press, Serdang (2009)
31. Morandim-Giannetti, A.A., Agnelli, J.A.M., Lancas, B., Magnabosco, R., Casarin, S.A., Bettini, S.H.P.: Lignin as additive in polypropylene/coir composites: Thermal, mechanical and morphological properties. Carbohydr. Polym. **87**, 2563–2568 (2012)
32. Mohamed, A.R.: Physical, Mechanical and Thermal Properties of Pineapple Leaf Fibers (PALF) and PALF-Reinforced Vinyl Ester Composites, Ph.D Thesis, Universiti Putra Malaysia, Serdang, Selangor, Malaysia (2010)
33. Mir, S.S., Nafsin, N., Hasan, M., Hasan, N., Hassan, A.: Improvement of physico-mechanical properties of coir polypropylene biocomposites by fibre chemical treatment. Mater. Des. **52**, 251–257 (2013)
34. Rashdi, A.A.A.: Moisture Absorption Capacity of Kenaf Fibre-Reinforced Unsaturated Polyester Composites and Its Effect on Their Mechanical Properties, Ph.D Thesis, Universiti Putra Malaysia, Serdang, Selangor, Malaysia (2010)
35. Rezali, K.A.M.: Mechanical Properties of Untraeted and Alkaline Treated-kenaf and Ramie-Fabric Reinforced Epoxy Composites, Master of Science Thesis, Universiti Putra Malaysia, Serdang, Selangor, Malaysia (2008)
36. Rozman, H.D., Ahmadhilmi, K.R., Abubakar, A.: Polyurethane (PU)—oil palm empty fruit bunch (EFB) composites: the effect of EFBG reinforcement in mat form and isocyanate treatment on the mechanical properties. Polym. Testing **23**, 559–565 (2004)
37. Russo, P., Carfagna, C., Cimino, F., Acierno, D., Persico, P.: Biodegradable composites reinforced with kenaf fibers: Thermal, mechanical, and morphological issues. Adv. Polym. Technol. **32**, 313–322 (2013)
38. Sahari, J.: Physio-Chemical and Mechanical Properties of Different Morphological Parts of Sugar Palm Fibre Reinforced Polyester Composites, Master of Science Thesis, Universiti Putra Malaysia, Serdang, Selangor, Malaysia (2011)

References

39. Sahari, J., Sapuan, S.M., Zainudin, E.S., Maleque, M.A.: Mechanical and thermal properties of environmentally friendly composites derived from sugar palm tree. Mater. Des. **49**, 285–289 (2013)
40. Sapuan, S.M., Harimi, M., Maleque, M.A.: Mechanical properties of epoxy/coconut shell filler particle composites. Arab. J. Sci. Eng. **28**, 171–181 (2003)
41. Sapuan, S.M., Zan, M.N.M., Zainudin, E.S., Arora, P.R.: Tensile and flexural strengths of coconut spathe-fibre reinforced epoxy composites. J. Trop. Agric. **43**, 63–65 (2005)
42. Sapuan, S.M., Leenie, A., Harimi, M., Beng, Y.K.: Mechanical properties of woven banana fibre reinforced epoxy composites. Mater. Des. **27**, 689–693
43. Sapuan, S.M., Siregar, J.P.: Mechanical properties of pineapple leaf fibre reinforced high impact polystyrene composites. In: Proceedings of the 20th Australasian Conference on the Mechanics of Structures and Materials, Toowoomba, Australia, 2–5 Dec 2008, pp. 295–299 (2008)
44. Sapuan, S.M., Pua, F., El-Shekeil, Y.A., AL-Oqla, F.N.: Mechanical properties of soil buried kenaf fibre reinforced thermoplastic polyurethane composites. Mater. Des. **50**, 467–470 (2013)
45. Shackelford, J.F.: Introduction to Materials Science and Engineers, 7th edn. Pearson Education Inc., Upper Saddle River (2009)
46. Siregar, J.P.: Tensile and Flexural Properties of Arenga Pinnata Filament (Ijuk Filament) Reinforced Epoxy Composites, Master of Science Thesis, Universiti Putra Malaysia, Serdang, Selangor, Malaysia (2005)
47. Siregar, J.P., Sapuan, S.M.: The effect of compatabilizing agents on mechanical properties of pineapple leaf fibre (PALF) reinforced high impact polystyrene composites. Int. J. Polym. Technol. **3**, 8185 (2011)
48. Siregar, J.P.: Effects of Selected Treatments on Properties of Pineapple Leaf Fibre Reinforced High Impact Polystyrene Composites, Ph.D Thesis, Universiti Putra Malaysia, Serdang, Selangor, Malaysia (2011)
49. Stael, G.C., Tavares, M.I.B., d'Almeida, J.R.M.: Impact behavior of sugarcane bagasse waste–EVA composites. Polym. Testing **20**, 869–872 (2001)
50. Threepopnatkul, P., Kaerkitcha, N., Athipongarporn, N.: Effect of surface treatment on performance of pineapple leaf fiber–polycarbonate composites. Compos. B Eng. **40**, 628–632 (2009)
51. Uma Devi, L., Bhagawan, S.S., Thomas, S.: Mechanical properties of pineapple leaf fiber reinforced polyester composites. J. Appl. Polym. Sci. **64**, 1739–1748 (1997)
52. Vallejos, M.E., Curvelo, A.A.S., Teixeira, E.M., Mendes, F.M., Carvalho, A.J.F., Felissia, F.E., et al.: Composite materials of thermoplastic starch and fibers from the ethanol–water fractionation of bagasse. Ind. Crops Prod. **33**, 739–746 (2011)
53. Venkateshwaran, N., Perumal, A.E., Alavudeen, A., Thiruchitrambalam, M.: Mechanical and water absorption behaviour of banana/sisal reinforced hybrid composites. Mater. Des. **32**, 4017–4021 (2011)
54. Wambua, P., Ivens, J., Verpoest, I.: Natural fibres: can they replace glass in the fibre reinforced plastics? Compos. Sci. Technol. **63**, 1259–1264 (2003)
55. Wirawan, R.: Thermo-Mechanical Properties of Sugarcane Bagasse-Filled Poly(vinyl Chloride) Composites, Ph.D Thesis, Universiti Putra Malaysia, Serdang, Selangor, Malaysia (2011)
56. Wirawan, R., Sapuan, S.M., Yunus, R., Abdan, K.: The effects of thermal history on tensile properties of poly(vinyl chloride) and its composite with sugarcane bagasse. J. Thermoplast. Compos. Mater. **24**, 567–579 (2011)
57. Xie, Y., Hill, C.A.S., Xiao, Z., Militz, H., Mai, C.: Silane coupling agents used for natural fiber/polymer composites: a review. Compos. A Appl. Sci. Manuf. **41**, 806–819 (2010)

58. Yusoff, M.Z.M.: Mechanical Properties of Oil Palm Fibre-Thermoset Composites, Master of Science Thesis, Universiti Putra Malaysia, Serdang, Selangor, Malaysia (2009)
59. Zainudin, E.S., Sapuan, S.M., Abdan, K., Mohamad, M.T.M.: Mechanical properties of compression moulded banana pseudo-stem filled unplasticized polyvinyl chloride (uPVC) composites. Polym. Plast. Technol. Eng. **48**, 97–101 (2008)

Chapter 5
Design and Materials Selection of Tropical Natural Fibre Composites

Abstract In this chapter, two main topics in product development of products from tropical natural fibre composites; product design and materials selection are discussed. Engineering design process is presented and the emphasis is on the total design. Conceptual design and detail design are key activities in design but only conceptual design is emphasized in this book. Design approach for natural fibre composites may be different from metals, conventional composites, polymer and ceramic. Various idea generation techniques are discussed. Concept evaluation techniques for tropical natural fibre composite product development were reviewed. Design for manufacture is briefly reviewed. Materials selection using various tools and techniques suitable selection of tropical natural fibre composite materials is finally presented. Topic of design of products from tropical natural fibre reinforced polymer composites is still new. A lot of areas within this topic that can be explored such as in the areas of design for manufacture of natural fibre composites, design analysis during conceptual and detail design of products from natural fibre composites, design for sustainability for natural fibre composites, mould flow analysis for natural fibre composites and design to cost for natural fibre composite products. All those topics were originally intended to included in this chapter but due to space constraints, they are proposed to be included in future publication.

Keywords Conceptual design · Materials selection · Design for manufacture · Concept evaluation · 'Green' design

5.1 Introduction

Product development is normally carried out in systematic manners in order to achieve products that can be marketed and at the same time they are developed at high quality for enhancing the quality of life of mankind. Research on design of tropical natural fibre composites is scarcely conducted. It may be because products made from tropical natural fibre composites are also scarcely found in the market.

© Springer Science+Business Media Singapore 2014
M.S. Salit, *Tropical Natural Fibre Composites*,
Engineering Materials, DOI 10.1007/978-981-287-155-8_5

Design and materials selection of tropical natural fibre composites are among the research activities carried out at Engineering Composites Research Group, Universiti Putra Malaysia. Normally, for design, conventional methods used in product design and composites product design can be adopted with slight modification. Similarly materials selection used in conventional materials and composites can be adopted for tropical natural fibre composites with some modifications. Engineering product designers and materials engineers are normally integral parts in composite product development team and it is accepted if we assume materials selection is a major activity in product design. In this chapter, then we talk about product design, design alone is not the only activity. It also relates to market investigation, information gathering, asking customers' need, developing product design specifications (PDS), materials consideration, supplier involvements, sustainability and manufacturing planning. Due to space limitation, only some of these activities will be presented in this chapter. Selected case studies of design of tropical natural fibre reinforced polymer composite components using kenaf, sugar palm, banana pseudo-stem and oil palm fibre composites are presented here. No work on design is discussed using pineapple leaf, coconut, and sugarcane fibre composites.

5.2 Engineering Design Process

Engineering design process is a process for designing a product from design brief, conceptual design, detail design and manufacture. Many design process models were developed by various experts such as Ertas and Jones [12], Dieter [10], Pahl et al. [32] and Pugh [33]. Pugh [33] proposed a total design (TD) method for product design and this method was very popular in the 1990s and still widely being used. The model starts with market investigation, PDS, conceptual design, detail design and manufacture. He did not include embodiment design as part of the model. However, Pahl et al. [32] introduced embodiment design after conceptual design and despite that fact that detail design is a very important activity in engineering design, in their book section of detail detail design is not included; may be because detail design involves many different disciplines of engineering that can distract the readers from appreciating 'design'. The design of composite product is largely dependent on the product's intended use. Decision must be made on how strong and stiff the product, how long it will be in use, the budget that is available, and what type of standard of finish [27].

5.3 Conceptual Design

Although, Pugh [33] regarded market investigation and product design specifications (PDS) as different activities to conceptual design and they were done before conceptual design, Dieter [10] held different view. Dieter [10] stated that

5.3 Conceptual Design

conceptual design includes recognition of need, definition of problem, gathering of information, developing design concepts and evaluation of concepts. By taking the statement of Dieter [10] as guides, market investigation and product design specification (PDS) are considered as parts of the activities of conceptual design in this chapter, although the author's earlier publication Sapuan and Maleque [40], Sapuan et al. [42], Sapuan [35, 37, 39] followed the model by Pugh [33].

5.3.1 Market Investigation

Information gathering is an important activity during market investigation stage. Pugh [33] has stated that the types of information needed; they can be in the forms of:

- Intellectual properties right (IPR) like patents and industrial design
- Reports, proceedings and reference books
- Manufacturers of competitive and analogous products
- Official and private representative bodies (standards)
- Statistical data
- Market data publications
- Filling in the gaps in recorded sources.

According to Hyman [19], there are two types of approaches in information gathering in product design. First, for a well developed area and the design in the area which not been done earlier. In the former, information is easy to find but in the latter, a designer may need to carry out experiments or develop new models for obtaining the information. Hyman [19] reported that for the design, information can be obtained from books, dictionaries, encyclopedias, handbooks, research journals, magazines, conference proceedings, technical reports, indexes, abstracts and reviews, code and standards, and patents. He further added that design information can be found professional societies and trade associations, Industrial research institutes and consortia, college and universities, government, personal network and continuing education courses.

Wright [53] discussed in his textbook methods for market research before a designer can carry out god design activities. Among the methods discussed include focus group meetings, mail questionnaires, telephone interviews, assessment of competitor products, internal company records, and market intelligence.

Misri [34] had performed information gathering through questionnaire (Fig. 5.1) before he can carry out the development of hybrid sugar palm/glass fibre reinforced polyester composite small boat.

5.3.2 Product Design Specification (PDS)

According to Suddin [49] a PDS is an important activity in product design because it sets the target to be achieved in the design. Mansor et al. [24], Davoodi et al. [9]

Fig. 5.1 Questionnaire form

		Excellent	Good	Average	Fair	Poor
Occupation	Race			Salary		Below RM1000
Age	Sex					RM1000~3000
						Above RM3000

		What is your favorite boat specifications	Excellent	Good	Average	Fair	Poor
1		Material type of boat you are interested in					
	a)	Fiberglass	☐	☐	☐	☐	☐
	b)	Wood	☐	☐	☐	☐	☐
2		The less weight boat is better	☐	☐	☐	☐	☐
3		The marginal safety is important in your boat	☐	☐	☐	☐	☐
4		Low maintenance costs	☐	☐	☐	☐	☐
5		Comfortable design	☐	☐	☐	☐	☐
6		More capacity size, the larger boats needed	☐	☐	☐	☐	☐

Comment :

Fig. 5.1 Questionnaire form

and Misri et al. [31] used PDS as design guides in the development of the products from natural fibre composites. Mansor et al. [24] developed PDS for hybrid glass-kenaf composites for automotive parking brake lever, Davoodi et al. [8] for hybrid glass-kenaf composites for automotive bumper beam and Misri et al. [31] for hybrid glass-sugar palm composites for small boat. From these studies, it is evidenced that tropical natural fibres can only be used by hybridizing with glass fibres for structural components due to the high strength requirements of he components. Pugh [33] had listed 32 elements of PDS for product design. Generally, a designer may not need to use all the elements of PDS and he/she chooses the PDS elements that are most relevant to his/her work. Mansor et al. [24] selected performance, weight, standards, disposal, environment and cost as important elements of PDS (Fig. 5.2) while Davoodi et al. [8] chose safety, size, maintenance, performance, installation, materials, weight, environment, manufacturing process, manufacturing cost and disposal as the important elements (Fig. 5.3) and the PDS elements used by Misri et al. [31] were performance, safety, materials, environment, manufacturing process, size and cost.

The PDS prepared by Misri et al. [31] for the design of a small boat using hybrid glass/sugar palm fibre reinforced composites is given as follows:

5.3 Conceptual Design

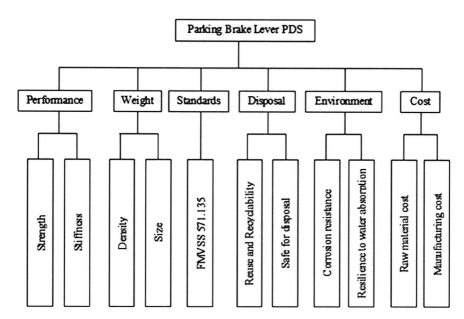

Fig. 5.2 PDS elements for the design of parking brake lever [24]

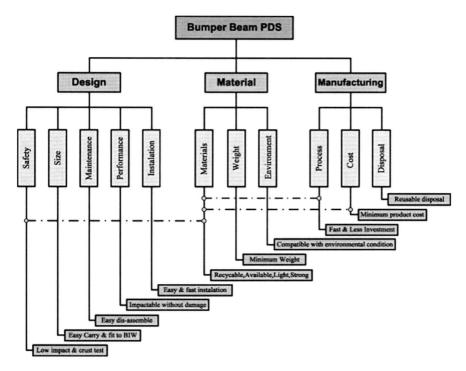

Fig. 5.3 PDS elements for the design of bumper beam [7]

A. Performance

- Generally, a composite boat will be used in lakes and small rivers.
- A composite boat requires a maximum number of 10 ribs.
- The maximum number of the passengers is usually four.

B. Safety

- A composite boat must be fitted with built-in float so that it will be floating even in an upside down condition.

C. Materials

The material used for the composite boat is sugar palm-glass fibre reinforced unsaturated polyester composite.

- must not be brittle.
- used must be easy to manufacture even for a complex shape to fulfill the needs of current boat style.
- must be easy to fabricate, does not rot or decay when exposed to the environment.

D. Environment

The composite boat

- will be used in lakes and small rivers.
- must be able to withstand all extreme weather conditions such as heat or rain when exposed to the outside environment.
- must perform and must not be damaged by temperature in the range of 10–50 °C.
- must have high durability, especially against pressure of river waves.

E. Manufacturing process

The manufacturing process of the composite boat must be

- on customer needs (based on orders).
- through manual process.

F. Size

- The maximum length of the composite boat must be 7 m.
- The maximum width of the composite boat must be 1.5 m.

G. Cost

- The maximum cost to manufacture the composite boat is RM 3,000*

  *One US dollar is equivalent to Malaysian Ringgit RM 3.33.

5.3.3 Concept Generation

Light weight is the reason for design with composites. In addition, natural fibre composites are chosen because of their low cost, biodegradability (as far as fibres are concerned), renewable and abundance. Therefore in designing products from natural fibre composites, those attributes must be kept in mind. Pahl et al. [32] have enumerated various solution finding methods during conceptual design stage. They divided solution finding methods into four parts namely conventional methods, intuitive methods, discursive methods and methods for combining solutions. Examples of conventional methods include information gathering, analysis of natural systems, analysis of existing technical systems, analogies and measurements and model tests. As far as intuitive methods, is concerned, Pahl et al. [32] have included brainstorming, method 635, gallery method, Delphi method, synectics and combination of methods. For discursive methods, they include systematic study of physical processes, systematic search with the help of classification schemes, and use of design catalogues. Finally for methods for combining solutions include systematic combination and combining with the help of mathematical methods. Perhaps the above-mentioned methods by Pahl et al. [32] are only some selected methods available.

Other design experts have come up with other methods such as Cross [7] and Ulrich and Eppinger [50]. It is reported in Ulrich and Eppinger [50] that some hints found that can generate concepts such as making analogies, wish and wonder, using related stimuli, using unrelated stimuli, setting quantitative goals and using the gallery method. It is not possible to go into detail on the individual methods mentioned above, but it is only selected methods are dealt in this chapter. Buzan [6] proposed a mind mapping technique for creativity and this technique is suitable for concept generation in product development of natural fibre composites. Note making exercise is a powerful tool to generate design concepts [41]. It helps to organize and clarify one's thinking and to see the 'whole picture' [6]. Figure 5.4 shows a preliminary mindmapping tree for the design of composite bumper system.

During idea generation stage, the use of sketches in explaining the ideas of designers is a very common practice. Figure 5.5 shows a sketch of idea for skateboard from natural fibre composites during an undergraduate course on the design of natural fibre composite products. Design idea can also be presented in a three dimensional from using solid modeling system as shown in Fig. 5.6.

Misri et al. [31] used many different conceptual design tools in the development of hybrid glass/sugar palm reinforced unsaturated polyester composites. It includes brainstorming, PDS, questionnaire survey, and morphological charts. Sapuan [37]

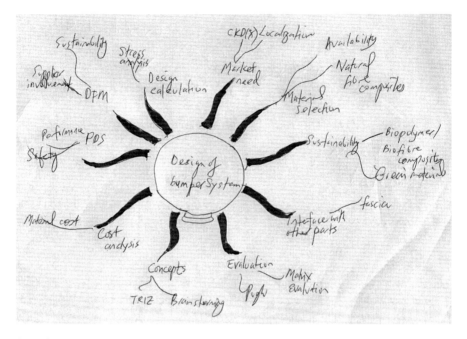

Fig. 5.4 Preliminary mindmapping sketch for the design of composite bumper system

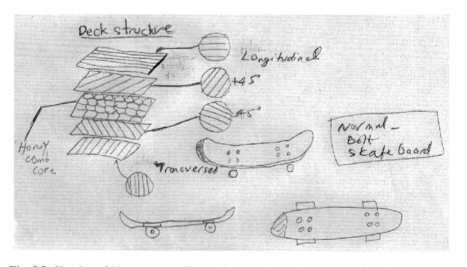

Fig. 5.5 Sketches of idea generation for the design of natural fibre composite skateboard

5.3 Conceptual Design

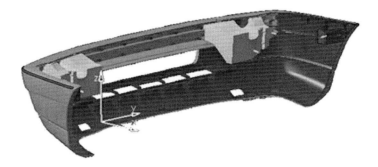

Fig. 5.6 A three dimensional representation of hybrid glass-kenaf fibre composite bumper beam [8]

also used methods like PDS, brainstorming, and matrix evaluation technique in the design of motor cycle pump from oil palm fibre reinforced epoxy composites. Sapuan and Maleque [40] (banana pseudo-stem telephone stand), Sapuan et al. [42] (banana pseudo-stem composite coffee table), Sapuan et al. [38] (banana pseudo-stem composite Holy Quran stand), Sapuan and Naqib [43] (banana pseudo-stem composite book shelf), Ham et al. [16] (oil palm composite telephone casing), Misri et al. [30] (sugar palm composite large table), Mazani et al. [25] (kenaf composite shoe rack) and Khamis et al. [22] (banana pseudo-stem composite chair) also used design concept generation techniques for a wide range of products.

One of the famous idea generation techniques is TRIZ [52]. TRIZ is a Russian acronym for "Teoriya Resheniya Izobreatatelskikh Zadatch" which is translated in English as Theory of Inventive Problem Solving. It was developed by a Russian patent Engineer named Genrich Altshuller. TRIZ was used by Mansor et al. [24] in conceptual design of hybrid glass-kenaf composite automotive parking brake lever. They integrated with technique with Analytical hierarchy Process (AHP) and morphological chart to come up with the best design concept. The TRIZ contradiction matrix and 40 inventive principles solution tools were used in the early design stage and the ideas generated were refined using morphology chart to come up with the design concepts for parking brake lever (see Fig. 5.7).

Misri et al. [31] also used morphological chart to design hybrid sugar palm/glass fibre reinforced composite boat. In this method, the design concepts of the boat were developed based on the combination of various sub-functions. The sub-functions include manufacturing process, complexity of boat, number of bones, number of passengers, safety, and place of use (lake, river and sea). Figure 5.8 shows morphology chart for the composite boat (concept 1). It has five solutions options to its sub function listed as shown in the Table to generate the concept of the boat. The concept can be used on sea, so it requires seven structural ribs, the complexity of the shape of the boat is medium, the fabrication process is hand lay-up and it can accommodate 7 passengers and two buoys.

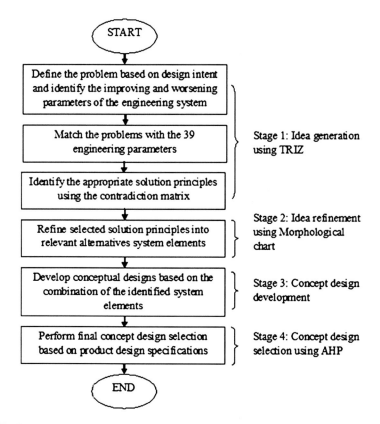

Fig. 5.7 Flow chart of integrated TRIZ, AHP and morphological chart techniques for concept generation [24]

5.3.4 Concept Evaluation

Weighted objective method [7] is the most popular method for design concept evaluation. It is also called the concept scoring matrix [50]. Sapuan and Maleque [40] and Sapuan et al. [42] used weighted objective method in the evaluation of design concepts for banana pseudostem woven fibre reinforced epoxy composite furniture namely telephone stand and multipurpose table respectively. The same method was used to evaluate the best design concept for motor cycle pump made from oil palm fibre reinforced polymer composites [37]. Misri et al. [31] used weighted objective method in evaluating the design concepts for glass-sugar palm composite boat as shown in Fig. 5.9.

In this method, design objectives are listed which comprise technical and economic considerations, as well as user requirements [7]. Then, the objectives are listed in a rank order of importance, i.e. relative 'weights' for the objectives are determined. In their study, Misri et al. [31] relative weightings of the objective

5.3 Conceptual Design

Sub Function	Solution				
	1	2	3	4	5
Capability on boat structure	no rib	2 ribs	3 ribs	5 ribs	7 ribs
Shape of boat	simple	medium	complex		
Boat fabrication process	hand lay-up	resin transfer moulding (RTM)	resin infusion moulding		
Number of passengers	1	2	3	5	7
Safety	1 buoys	2 buoys	3 buoys	4 buoys	5 buoys
Boat application	lake and small river	river	Sea		

Fig. 5.8 Morphological chart for the composite boat (concept 1)

Objective	Weight	Concept-1		Concept-2		Concept-3		Concept-4		Concept-5	
		Score	Value	Score	Value	Score	Value	Score	Value	Score	Value
Low cost	0.2	5	1.00	8	1.60	7	1.40	5	1.00	7	1.40
Light weight	0.2	4	0.80	8	1.60	7	1.40	7	1.40	7	1.40
Manufacturability	0.25	6	1.50	7	1.75	8	2.00	5	1.25	8	2.00
Material	0.25	7	1.75	6	1.50	8	2.00	6	1.50	8	2.00
Appearance	0.1	8	0.80	6	0.60	8	0.80	8	0.80	6	0.60
Overall utility value	1.00	5.85		7.05		7.60		5.95		5.40	

Fig. 5.9 Weighted objective method for evaluation of concepts for the composite boat

are assigned by sharing a certain number of number, in this case 1.0, among all the objectives. When the total of 1.0 is allocated amongst all five objectives, then the summation of weightings of all objectives must be 1 (i.e. $0.2 + 0.2 + 0.25 + 0.25 + 0.1 = 1.00$). Then the designer must establish utility scores for each of the objectives. A point scale from 4 to 8 was used. Finally, the score is multiplied by its weighted value for each concept; the concept with the highest overall utility value of 7.6 is selected as the best concept (i.e. concept 3).

Pugh evaluation method [33] is another popular concept evaluation method widely used to select the best concept based on comparison of candidate designs and existing design normally referred to as Datum. Each concept was evaluated based on several design requirements and the performance of each concept is compared with Datum. If the concept is better than Datum in fulfilling the design requirement, the score '+' is assigned. If it is inferior to Datum, it is assigned the score '−'. If it performed the same as Datum, 'S' is assigned. For the concept with the highest positive difference between '+' and '−', then this concept is selected as the best concept. Figure 5.10 shows the Pugh evaluation method used to evaluate the design concept of a table made from sugar palm fibre reinforced polymer composites. In this design, concept six is selected as the best concept because it scores the highest difference between + and −, i.e. $6−1 = 5$.

Apart from these two conventional methods for concept evaluation, Mansor et al. [24] and Davoodi et al. [8] used AHP and TOPSIS methods to evaluation the design concepts for hybrid glass/kenaf polymer composite and parking brake lever automotive bumper beam respectively. Mansor et al. [24] used AHP to select the best concept of brake lever. After five design concepts of the component were developed using TRIZ and morphological chart (see Fig. 5.11), the AHP method was used to perform the multi-criteria decision making process to select the best concept for polymer composite automotive parking brake lever component. AHP is a decision making tool which can be used to carry out decision task by decomposing a complex problem into a multi-level hierarchical structure of objectives, criteria, sub-criteria, and alternatives [34]. In this study, concept design selection was performed using AHP commercial package called Expert Choice. AHP hierarchy frame work listed the goal as selection of concept for automotive parking brake lever (level 1). The main criteria were performance, weight and cost (level 2). There are six sub-criteria selected i.e. stress, deformation, mass volume, raw material cost and shape complexity (level 3). For level 4; i.e. alternatives; are represented by five design concepts. In this technique, pair-wise comparison was used to judge all the design concepts with respect to each main criterion and sub-criterion. The overall AHP results of the concept design selection are shown in Fig. 5.12, which gave that concept 2, with the highest score is selected as the best design. Similar work on design concept selection was performed [17] for automotive composite bumper beam. But bumper beam designs considered were not made from natural fibre composites.

Davoodi et al. [8] used TOPSIS method to evaluation design concept for bumper beam. In his study, rather than evaluating the design concepts in the complete form of product, they evaluated few geometries of bumper beam which

5.3 Conceptual Design

Concept		First	Second	Third	Fourth	Fifth	sixth
Easy Manufacture	D	I	-	–			I
Mass		I	-	+			+
Effeciacy Cost	A	+	-	I	-		+
Ergonomic	T	-	+	I	-		-
Strength		-	+	-	S		+
Less Maintanance	U	S	-	-	S		+
Stability		-	+	I	S		+
Apperance	M	S	I	+	-		S
Total Score		Σ+ 4	4	6	4		6
		Σ- 2	4	2	1		1
		ΣS 2	0	0	3		1

Fig. 5.10 Pugh evaluation method to select the best concept for sugar palm composite table

called these geometries as concepts as well. All the concepts of the bumper beams shared some parameters in common: frontal curvature, thickness, and overall dimensions. The design objectives of the study were deflection, strain energy, mass, cost, easy manufacturing, and the possibility of including rib. They were placed in the evaluation matrix. The selection of the best const concept was carried out using TOPSIS method. TOPSIS stands for the Technique for Order of Preference by Similarity to Ideal Solution. It is a multi-criteria decision analysis method initially introduced by Hwang and Yoon [18] based on the idea that the best alternative should have the shortest distance from an ideal solution. The algorithm considers ideal and non-ideal solution it help designer to evaluate ranking and select the best option. From the evaluation, double hat profile (DHP) was selected as the best concept and it was refined further in the later stage of the design. Figure 5.13 shows the concept evaluation of automotive bumper beam using TOPSIS method.

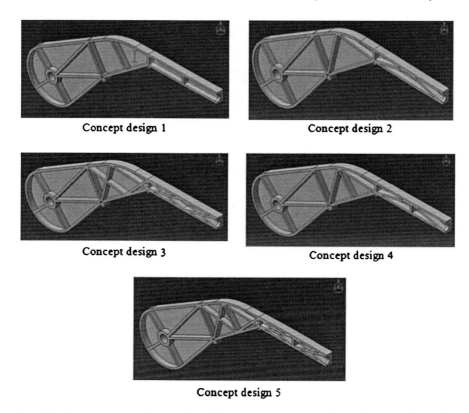

Fig. 5.11 Design concepts for glass/kenaf fibre composite automotive parking brake lever [24]

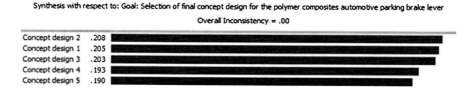

Fig. 5.12 Overall AHP results of the concept design selection [24]

5.4 Detail Design

Dieter [10] and Pahl et al. [32] have mentioned the next stage after conceptual design is embodiment design and a chapter on detail design in earlier edition of the book of Pahl et al. [32] was removed from the new edition of the book. It is not consistent with what was proposed by Pugh [33] where he stated that the next stage after conceptual design is detail design. Dieter [10] differentiate between

5.4 Detail Design

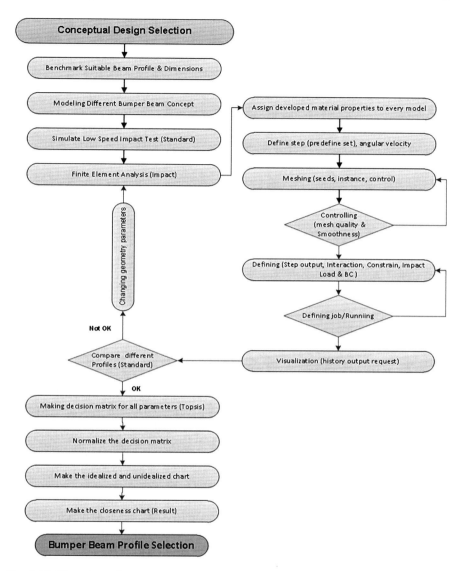

Fig. 5.13 Flow chart of concept evaluation of automotive bumper beam using TOPSIS method [8]

embodiment design and detail in the following manner. In embodiment design, a structured development of the design concept is carried out. In this stage, decisions on strength, material selection, size, shape, and spatial compatibility are made. Embodiment design tasks include product architecture, configuration design of parts and components and parametric design of parts and components [10]. In detail design, 'the design is brought to the stage of a complete engineering description of a tested and producible product' [10]. In this stage, specifications for each custom and standard part are made available, detailed engineering drawings generated by computers are prepared, and design review is carried out. Figures 5.14 and 5.15 show the engineering drawing of banana pseudo-stem fibre reinforced epoxy composite household telephone stand and hybrid glass/sugar palm fibre reinforced unsaturated polyester composite boat respectively. However, in this chapter, majority of the discussion is related only to conceptual design.

5.5 Design for Manufacture

In the past, product development was carried out in isolation; a designer designed a product without any communication with other staff within the organization. In this environment, once the detail was complete, it was handed over to manufacturing engineer who was responsible for its fabrication. This syndrome is called 'throw over the wall' and there is a departmental barrier between design and manufacture. Due to stiff competition in the marketplace, products cannot be developed in isolation and have to be developed by considering all manufacturing related issues such as assembly, cost and tolerances. As a result, products are developed with higher quality, shorter time and at lower cost. Therefore, there is was real need to seek for methods and tools that would enable industries to carry out design and development work by considering all manufacturing related issues early in the design process. The obvious method to choose if design for manufacture (DFM) or concurrent engineering (CE). Mazumdar [26] defined DFM as 'a practice for designing products, keeping manufacturing in mind'. Mazumdar [26] relates DFM to total design method developed by Pugh [33] where the activities like concept generation and concept evaluation are found to in common in both method. In addition, DFM promotes, among others, narrowing design selection to optimum design, minimizing cycle time and cost, targeting for high quality product and minimizing the number of parts and assembly time.

Sapuan and Mansor [48] carried out a review of concurrent engineering approach in developing products from composites. This work is part of the research carried out by the first author during his sabbatical leave in 2013–2014 in Universiti Malaya, Kuala Lumpur entitled manufacturability of natural fibre composites. They emphasized that CE can be applied in tropical natural fibre composite product development. Sapuan and Mansor [48] have also pointed out, the most convenient way to understand the CE principle in composite product development is by associating it with the Pugh total design model or any established design process model. It is because of the abstract nature of CE philosophy and implementing it with total design model revealed the practical side of it.

In actual fact, the work on the implementation of CE in the development of products from natural fibre composites is very limited. Sapuan [44] in his professorial inaugural lecture had reported some work on CE of selected products

5.5 Design for Manufacture

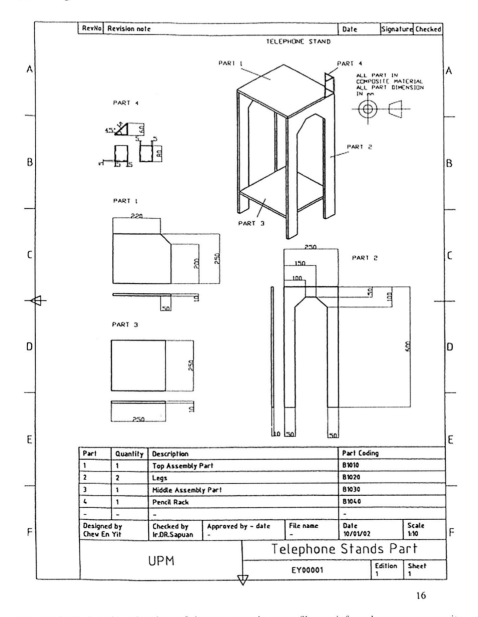

Fig. 5.14 Engineering drawing of banana pseudo-stem fibre reinforced epoxy composite household telephone stand [40]

made from tropical natural fibre composites. He reported various work on the use CE in natural fibre composite product development and among the products reported include chair, table, computer casing, trolley, and small boat. In all these

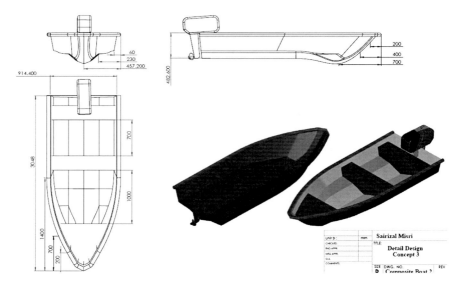

Fig. 5.15 Engineering drawing of hubrid glass/sugar palm fibre reinforced unsaturated polyester composite boat [31]

development, CE tools used were mainly associated with Pugh total design model. The tropical natural fibres used include kenaf, oil palm, sugar palm and banana pseudo-stem.

5.6 Materials Selection in Design

Materials selection is an important activity and integral part of engineering product design. Selection of the most suitable material can highly influence the performance of the design. The difficulty in the selection of materials depends on the complexity of the product being designed. Designer should have sufficient knowledge on technical as well as economic aspects of the design before he/she is able to decide on which material that is suitable for intended application. This is because inappropriate selection of materials may lead to heavy financial losses as depicted in Fig. 5.16 and materials selection must be properly managed.

Materials selection is a process of selecting the most optimum materials for specific purposes after satisfying all the requirements. It is a process normally performed by design engineer or materials engineer in the product development process. The purpose of materials selection is [15]:

> the identification of materials, which after appropriate manufacturing operation, give the dimensions, shape and properties necessary for the end product or component to perform its required function at the lowest possible total manufacturing cost.

5.6 Materials Selection in Design

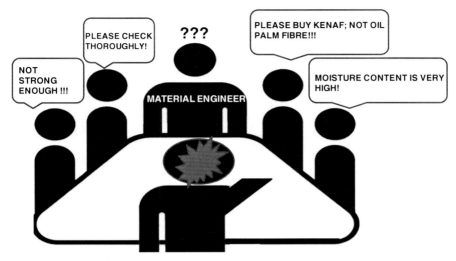

Fig. 5.16 The need to manage materials selection

According to Ashby [4] in product design process, from concept to detail design, different design tools and material data are required. During conceptual design stage, design tools are normally in the form of function modeling, viability studies and approximate analysis, during embodiment design, geometric modeling, simulation modellings and cost modeling are important and during detail design stage, component modeling and finite element modeling are important. As for the materials selection, during conceptual design, it requires data for all materials, during embodiment design, data for a subset of materials and in detail design, data for one material are need (Fig. 5.17).

Materials selection is an activity that is related to many other activities and in materials selection, Pahl et al. [32] have stressed that the knowledge of various different fields such as layout and form design, safety, ergonomics, production, quality control, assembly, transport, operation, maintenance, costs, scheduling and recycling are important. [3] pointed out that in the selection of materials, inputs from engineering analysis and synthesis must be considered along with inputs from industrial design such as colour, texture, form, shape and aesthetic.

Traditionally materials selection was performed manually with the help of materials handbooks or materials brochures. These processes were only effective when limited number of candidate materials available. When number of candidate materials is becoming very large, materials selection is normally performed with aid of computer software system. Most of the time, many computer-based materials selection systems are available to assist the designer in the selection of the most optimum materials. Database like Cambridge Engineering Selector is the most popular and well accepted computer aided materials selection system.

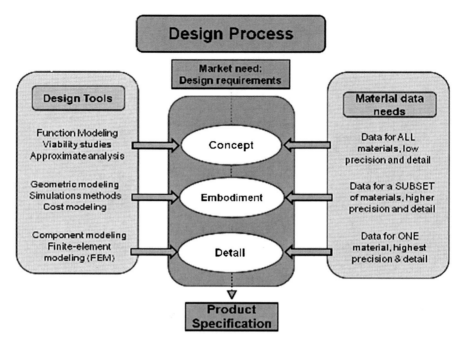

Fig. 5.17 The design model with the tools and materials data needed [4]

However, for the purpose of research and development, many techniques have been developed for selecting materials particularly employing artificial intelligent techniques such as expert system, fuzzy logic, artificial neural network, genetic algorithm, and analytical hierarchy process (AHP). Many multi-criteria decision making (MCDM) techniques are used as selection tools in the recent years for materials selection.

Hambali [17] used AHP as a MCDM tool to perform materials selection task for automotive bumper beam from composite materials. Bumper beam has the function to absorb the bulk energy and provides protection to the rest of vehicle [5]. In this study, materials selection was performed using Expert Choice software. The goal of the AHP frame work is for materials selection of automotive bumper beam (level 1). The main criteria selected were energy absorption, performance, cost, weight, service conditions, manufacturing process, environmental consideration and availability (level 2). The sub-criteria selected include impact toughness, flexural strength, flexural modulus, raw material cost, low density, resistance to corrosion, water absorption, shape, recyclability, disposal, availability of raw materials and availability of materials information (level 3). For level 4; i.e. alternatives; are represented by six different types of composite materials. In this technique, pair-wise comparison was used to perform judgements on all the candidate materials with respect to each main criterion and sub-criterion.

5.6 Materials Selection in Design

Jahan [21] carried out materials selection using an interesting tool called VI-KOR method. VIKOR method can be used to rank and to select the optimum material. It placed an emphasis on ranking and selecting the alternatives with conflicting criteria and with different units. In VIKOR method, the compromised ranking is carried out by comparing the measure of closeness to the ideal alternative, and compromise means an agreement established by mutual concessions. A literature review [20] based on this work is highly cited in major databases (61 citations in Scopus from 2010 to January 2014).

Fairuz [13] and Fairuz et al. [14] have performed materials selection for engineering products made from composite materials and the selection process was carried out implementing expert system approach using IF/THEN logic rules. Various criteria and attributes such as tensile strength, tensile modulus, flexural strength, flexural modulus, impact strength, density, and water absorption of polymer based composites were considered. The rule-based system in Logic Block describes the rules as units. In selecting the material, knowledge related to dependencies among fact are required. Such dependencies in the Logic Block system have general form of rule: If *<premise>* Then *<conclusion>*, if the premises are true, then the conclusions should also be true.

First rule: If *premise* A Then *conclusion* A
Second rule: If *premise* B Then *conclusion* B
Third rule: If *premise* C Then *conclusion* C

Sapuan and Abdalla [36] performed materials selection for automotive pedal box system using an expert system similar to the work of Fairuz [13] and Fairuz et al. [14]. In their study, commercial expert system shell called KEE was used to perform materials selection. Over thirty candidate composite materials were made available for the selection and they were mainly taken from materials handbooks and materials databases like CAMPUS and FUNDUS. Rule based system implementing IF-THEN was used to perform forward chaining process and the materials data were stored in frame-based system which comprised all the important data on materials properties. Two main materials selection drivers; Young's modulus and tensile strength were also parts of the database system. The major components of a pedal box system include mounting bracket, accelerator, clutch and brake pedals and the selection of materials for these components were performed.

5.7 Materials Selection for Tropical Natural Fibre Composites

Materials selection is an established area of research [3, 4] and materials selection for natural fibre composites is completely a new field of research and it is a difficult task to perform as materials data and references related to these materials are very limited. In addition, it is very difficult to separate between the materials selection of natural fibre composites with the rest of materials. This chapter is concerned with the materials selection process of engineering products from tropical natural fibre composites. Previous works that were dealing with this topic are very limited.

As with other composite materials, materials selection of natural fibre composite materials is a challenging task, because of the anisotropic nature of the materials that required the tailored made consideration of products and materials. Ermolaeva et al. [11] developed an optimization system for material selection of some engineering materials including natural fibre composites. Because of the vast variation in matrix type, fibre type, fibre arrangement, fibre dimension, manufacturing method, materials selection that involved natural fibre composites in particular and of composite materials in general is a daunting task to execute.

Miller et al. [28] developed multi-criteria materials selection system for natural fibre composites. The selection system focused on selecting various different types of reinforcing fibres for poly(b-hydroxybutyrate)-co-(b-hydroxyvalerate) composites. The materials selection system was applied to weigh properties with certain criteria that were environmentally related namely greenhouse gas emissions, demand for fossil fuel, and Eco-Indicator '99 score.

Sapuan and Mujtaba [45] developed a materials selection system using artificial neural network (ANN) for various types of natural fibre composites. Data were collected on the information from published papers. A total of 121 datasets for 121 different natural fibre composites were compiled. The datasets were different in terms of combinations of fibre loadings, types of fibres, fibre arrangements, and fibre treatment agents and they were used in the ANN system. Three main challenges in collecting comparable data include incomplete dataset, the use of fibre volume versus fibre weight and missing some of important information. Materials selection was performed in two stages. Stage one was concerned with ANN while stage two was using multi-attribute decision making to arrive at the best solution.

Sapuan et al. [46] used analytic hierarchy process (AHP) to perform materials selection to select the most suitable material for automotive dashboard panel. Database of properties of natural fibre composites composed data on density, Young's modulus and tensile strength. There are 29 types of natural fiber composites were considered as materials candidates in the selection process. Then judgement was carried out using pair-wise comparison followed by synthesizing pair-wise judgements and calculating priority vectors. Then consistency analysis was determined using consistency ratio. The appropriate natural fibre polymer composite for dashboard was selected. Finally sensitivity analysis was used to verify the results.

Mansor et al. [23] performed materials selection study on automotive brake parking lever (Fig. 5.18) using analytical hierarchy process. The candidate materials for this product are natural fibres and these fibres are intended to be hybridized with glass fibre. The initial stage in the materials selection process was the development AHP hierarchical framework. Expert Choice 11 software was used in this study. The best natural fibre composite for automotive parking brake lever was obtained by considering main criteria and sub-criteria in the AHP hierarchy system (Fig. 5.19) Then pair-wise comparison between the candidate

5.7 Materials Selection for Tropical Natural Fibre Composites

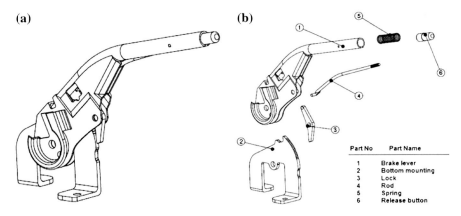

Fig. 5.18 3D CAD model of a commercial passenger vehicle center lever parking brake design in **a** assembly view, and **b** exploded view [23]

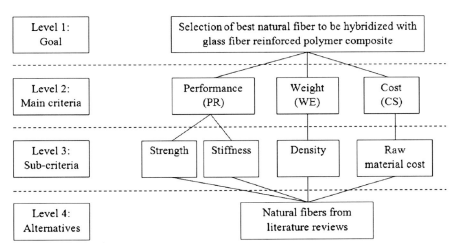

Fig. 5.19 Hierarchy flow chart in materials selection of hybrid glass/kenaf fibre composite automotive parking brake lever [23]

materials was performed as shown in Fig. 5.20. Sensitivity analysis was also performed to verify the study.

Sapuan et al. [47] developed a natural fibre composite material selection process using an expert system. The software used in the development of this expert system is PHPMyAdmin. The system was implemented to select the most suitable material for automotive dashboard. The system was written in Hypertext Preprocessor (PHP) language suitable for handling the administration of My Structure Query Language (MySQL) server in the World Wide Web. Various natural fibre reinforced polymer composite properties were considered to develop the system.

Kenaf bast fiber

Compare the relative preference with respect to: \ Strength

Oil palm EFB fiber

	Kenaf bast	Oil palm EF	Flax fiber	Hemp fiber	Jute fiber	Ramie fibe	Coir fiber	Sisal fiber	Cotton fibe	Bagasse fi	Pineapple
Kenaf bast fiber		3.75	1.61	1.03	1.16	1.86	4.23	1.33	2.33	3.21	1.75
Oil palm EFB fiber			6.05	3.63	3.23	2.02	1.13	2.82	1.61	1.17	6.56
Flax fiber				1.67	1.89	3.0	6.82	2.14	3.75	5.17	1.08
Hemp fiber					1.13	1.8	4.09	1.29	2.25	3.1	1.81
Jute fiber						1.6	3.64	1.14	2.0	2.76	2.03
Ramie fiber							2.27	1.4	1.25	1.72	3.25
Coir fiber								3.18	1.82	1.32	7.4
Sisal fiber									1.75	2.41	2.32
Cotton fiber										1.38	4.07
Bagasse fiber											5.61
Pineapple leaf fiber											
Banana fiber											
Sugar palm fiber	Incon: 0.00										

Fig. 5.20 Pair-wise comparison between (i) kenaf bast fiber and oil palm EFB fiber, and (ii) oil palm EFB fiber and ramie fiber, with respect to the tensile strength mechanical property for strength sub-criteria [23]

The main properties stored in the database include density, Young's modulus and tensile strength. The use of PHPMyAdmin software for natural fibre reinforced polymer composite materials selection process can assist designers to determine the most suitable natural fibre composites for engineering applications.

Yahaya et al. [51] carried out materials selection task for the selection of the best natural fibre as reinforcement in polymer composites. Analytical hierarchy process (AHP) method was used to evaluate a number of alternatives based on pre-defined criteria. This natural fibre will be hybridized with Kevlar 29 as fibres in laminated composites for vehicle spall liners. The fibre selected was kenaf fibre.

Ahmed Ali et al. [1] listed that there are eight steps to be performed in materials selection using AHP and they are:

1. problem definition
2. development of a hierarchy framework
3. construction of pair-wise comparison
4. performing judgement of pair-wise comparison
5. synthesis of pair-wise comparison
6. consistency analysis
7. checking consistency ratio (below 0.1)
8. development of priority ranking.

They claimed that the database of natural fibre composites are still do not meet today's requirements and data on natural fibres need to be collected from published work. They performed priority vector study and performed sensitivity analysis for the selection of materials for automotive door panel.

Ahmed Ali et al. [2] developed a Java based expert system for the selection of natural fibre reinforced polymer composites. The expert system for material selection was developed using the software programming language NetBean

JAVA™. The data were stored in relational database management system (RDBMS) software MySQL server database. The expert system was tested by means of a case study; automotive door panel. In the system, weighted-range method (WRM) was used to identify the range of value and to scrutinize the candidate materials. Java is platform independent and can be deployed in web based application and can be accessed through World Wide Web (www) so it can be used by multiple users internationally.

References

1. Ahmed, Ali, B.A., Sapuan, S.M., Zainudin, E.S., Othman, M.: Optimum materials selection using Analytical Hierarchy Process for polymer composites. In: Proceedings of the 2nd UPM-UniKL Symposium on Polymeric Materials 2013, Kuala Lumpur, 28th Feb 2013, pp. 76–81 (2013a)
2. Ahmed Ali, B.A., Sapuan, S.M., Zainudin, E.S., Othman, M.: Java based expert system for selection of natural fibre composite materials. J. Food Agric. Environ. **11**, 1871–1877 (2013)
3. Ashby, M., Johnson, K.: Materials and Design: The Art and Science of Material Selection in Product Design. Elsevier Butterworth-Heinemann, Amsterdam (2002)
4. Ashby, M.F.: Materials Selection in Mechanical Design, 3rd edn. Elsevier Butterworth-Heinemann, Amsterdam (2005)
5. Bernert, W., Bulych, S., Cran, J., Agle, D., Henseleit, K., et al.: Steel Bumper Systems for Passenger Cars and Light Trucks, Revision number 3, American Iron and Steel Institute, 30 June. http://bumper.autosteel.org (2006)
6. Buzan, T.: How to Mind Map®. Thorsons, London (2002)
7. Cross, N.: Engineering Design Methods, Strategies for Product Design, 2nd edn. Wiley, Chichester (1994)
8. Davoodi, M.M., Sapuan, S.M., Ahmad, D., Aidy, A., Khalina, A., Jonoobi, M.: Concept selection of car bumper beam with developed hybrid bio-composite material. Mater. Des. **32**, 4857–4865 (2011)
9. Davoodi, M.M.: Development of Thermoplastic Toughened Hybrid Kenaf/Glass Fibre-Reinforced Epoxy Composite for Automotive Bumper Beam, Ph.D Thesis, Universiti Putra Malaysia, Serdang, Selangor, Malaysia (2012)
10. Dieter, G.E.: Engineering Design: A Materials and Processing Approach, 3rd edn. McGraw-Hill, New York (2000)
11. Ermolaeva, N.S., Kaveline, K.G., Spoormaker, J.L.: Materials selection combined with optimal structural design: concept and some results. Mater. Des. **23**, 459–470 (2002)
12. Ertas, A., Jones, J.C.: The Engineering Design Process, 2nd edn. Wiley, New York (1996)
13. Fairuz, A.M.: Development of Material Selection Expert System for Polymer-Based Composite Materials, Master of Science Thesis, Universiti Putra Malaysia, Serdang, Selangor, Malaysia (2011)
14. Fairuz, A.M., Sapuan, S.M., Zainudin, E.S.: prototype expert system for material selection of polymeric-based composites for small fishing boat components. J. Food Agric. Environ. **10**, 1543–1549 (2012)
15. Gutteridge, P.A., Waterman, N.A.: Computer-aided materials selection. In: Bever, M.B. (ed.) Encyclopedia of Materials Science and Engineering, pp. 767–770. Pergamon Press, Oxford (1986)
16. Ham, K.W., Sapuan, S.M., Zuhri, M.Y.M., Baharuddin, B.T.H.T.: Development of computer casing using oil palm fibre reinforced epoxy composites. In: Proceedings of the 9th National Symposium on Polymeric Materials, 14–16 Dec 2009, Putrajaya, Malaysia (2009)

17. Hambali, A.: Selection of Conceptual Design using Analytical Hierarchical Process for Automotive Bumper Beam under Concurrent Engineering Environment, Ph.D Thesis, Universiti Putra Malaysia, Serdang, Selangor, Malaysia (2009)
18. Hwang, C., Yoon, K.: Multi Attribute Decision Making: Methods and Applications—A State-of-the-Art Survey. Springer, Berlin (1981)
19. Hyman, B.: Fundamentals of Engineering Design, 2nd edn. Pearson Education Inc., Upper Saddle River (1998)
20. Jahan, A., Ismail, M.Y., Sapuan, S.M., Mustapha, F.: Material screening and choosing methods—a review. Mater. Des. **31**, 696–705 (2010)
21. Jahan, A.: Improving Multi-Criteria Decision Making Algorithm for Material Selection, Ph.D Thesis, Universiti Putra Malaysia, Serdang, Selangor, Malaysia (2010)
22. Khamis, M.I.H., Sapuan, S.M., Zainudin, E.S., Wirawan, R., Baharuddin, B.T.H.T.: Design and fabrication of chair from hybrid banana pseudo-stem/glass fibre reinforced polyester composite. In: Proceedings of the 9th National Symposium on Polymeric Materials, 14–16 Dec, Putrajaya, Malaysia (2009)
23. Mansor, M.R., Sapuan, S.M., Zainudin, E.S., Nuraini, A.A., Hambali, A.: Hybrid natural and glass fibers reinforced polymer composites material selection using Analytical Hierarchy Process for automotive brake lever design. Mater. Des. **51**, 484–492 (2013)
24. Mansor, M.R., Sapuan, S.M., Zainudin, E.S., Nuraini, A.A., Hambali, A.: Conceptual design of kenaf fiber polymer composite automotive parking brake lever using integrated TRIZ–Morphological Chart-Analytic Hierarchy Process method. Mater. Des. **54**, 473–482 (2014)
25. Mazani, N., Sapuan, S.M., Ishak, M.R., Sahari, J.: Design and fabrication of kenaf fibre reinforced unsaturated polyester composite shoe rack. In: Proceedings of the UPM-UniKL Symposium on Polymeric Materials and the 3rd Postgraduate Seminar on Natural Fibre Composite 2012, Alor Gajah, Melaka, 2nd Feb, pp. 106–111 (2012)
26. Mazumdar, S.K.: Composites Manufacturing: Materials, Product, and Process Engineering. CRC Press, Boca Raton (2002)
27. McBeath, S.: Competition Car Composites: A Practical Handbook. Haynes Publishing, Somerset (2000)
28. Miller, S.A., Lepech, M.D., Billington, S.L.: Application of multi-criteria material selection techniques to constituent refinement in biobased composites. Mater. Des. **52**, 1043–1051 (2013)
29. Misri, S.: Design and Fabrication of Small Boat by Using Sugar Palm Fibre Reinforced Epoxy Composites, Master of Science Thesis, Universiti Putra Malaysia, Serdang, Selangor, Malaysia (2011)
30. Misri, S., Sapuan, S.M., Ishak, M.R., Sahari, J., Haron, M.: Design and fabrication of sugar palm fibre reinforced unsaturated polyester composite table. In: Proceedings of the UPM-UniKL Symposium on Polymeric Materials and the 3rd Postgraduate Seminar on Natural Fibre Composite 2012, Alor Gajah, Melaka, 2nd Feb, pp. 36–43 (2012)
31. Misri, S., Leman, Z., Sapuan, S.M.: Total design of a small boat using woven glass- sugar palm fibre reinforced unsaturated polyester composite. In: Sapuan S.M. (ed.) Engineering Composites: Properties and Applications. UPM Press,Serdang, pp. 224–47 (2014)
32. Pahl, G., Beitz, W., Feldhusen, J., Grote, K.H.: Engineering Design: A systematic Approach, 3rd edn. Springer, London (2007)
33. Pugh, S.: Total Design, Integrated Methods for Successful Product Engineering. Addison-Wesley Publishers Ltd., Workingham (1991)
34. Saaty, T.L.: Fundamental of Decision Making and Priority: Theory with the Analytical Hierarchy Process. RWS Publications, Pittsburg (2000)
35. Sapuan, S.M.: A Concurrent Engineering Design System for Polymeric-Based Composite Automotive Components, Ph.D. Thesis, De Montfort University, Leicester, UK (1998)
36. Sapuan, S.M., Abdalla, H.S.: A prototype knowledge-based system for the material selection of polymeric-based composites for automotive components. Compos. A Appl. Sci. Manuf. **29A**, 731–742 (1998)

References 101

37. Sapuan, S.M.: Design and fabrication of motorcycle pump from oil palm fibre reinforced epoxy composite. In: Presented at the 17th Symposium of Malaysian Chemical Engineers (SOMChE 2003), Penang, Malaysia, 29–30th Dec (2003)
38. Sapuan, S.M., Hassan, M.Y., Ismail, N., Johari, M.: Development of polymeric-based composite rehal. Al Nahdah Journal of RISEAP **22**, 37–38 (2003)
39. Sapuan, S.M.: Concurrent design and manufacturing process of automotive composite components. Assembly Autom. **25**, 146–152 (2005)
40. Sapuan, S.M., Maleque, M.A.: Design and fabrication of natural woven fabric reinforced epoxy composite for household telephone stand. Mater. Des. **26**, 65–71 (2005)
41. Sapuan, S.M., Maleque, M.A., Hameedullah, M., Suddin, M.N., Ismail, N.: A note on the conceptual design of polymeric composite automotive bumper system. J. Mater. Process. Technol. **159**, 145–151 (2005)
42. Sapuan, S.M., Harun, N., Abbas, K.A.: Design and fabrication of a multipurpose table using a composite of epoxy and banana pseudostem fibres. J. Trop. Agric. **45**, 66–68 (2007)
43. Sapuan, S.M., Naqib, M.: Design and fabrication of unidirectional pseudo-stem banana fibre reinforced epoxy composite small book shelf. Mech. Eng. Res. J. **6**, 1–7 (2008)
44. Sapuan, S.M.: Concurrent Engineering for Composites. UPM Press, Serdang, Selangor (2010)
45. Sapuan, S.M., Mujtaba, I.M.: Development of a prototype computational framework for selection of natural fiber-reinforced polymer composite materials using neural network. In: Sapuan, S.M., Mujtaba, I.M. (eds.) Composite Materials Technology: Neural Network Application, pp. 317–339. CRC Press, Boca Raton (2010)
46. Sapuan, S.M., Kho, J.Y., Zainudin, E.S., Leman, Z., Ahmed Ali, B.A., Hambali, A.: Material selection for natural fibre reinforced polymer composites using analytical hierarchy process. Ind. J. Eng. Mater. Sci. **18**, 255–267 (2011)
47. Sapuan, S.M., Mun, N.K., Hambali, A., Lok, H.Y., Fairuz, A.M., Ishak, M.R.: Prototype expert system for material system of polymeric composite automotive dashboard. Int. J. Phys. Sci. **6**, 5988–5995 (2011)
48. Sapuan, S.M., Mansor, M.R.: Concurrent engineering approach in the development of composite products: A review. Mater. Des. (in press) (2014)
49. Suddin, M.N.: Understanding the Nature of Specification Changes and Feedback to the Specification Development Process. Ph.D Thesis, Denmark Tekniske Universitet, Kgs. Lyngby, Denmark (2012)
50. Ulrich, K.T., Eppingher, S.D.: Product Design and Development, 3rd edn. McGraw-Hill, New York (2004)
51. Yahaya, R., Sapuan, S.M., Leman, Z., Zainudin, E.S.: Selection of natural fiber for hybrid laminated composite vehicle spall liners. In: Presented at International Conference on Advances in Mechanical and Manufacturing Engineering (ICAM^{2}E), Kuala Lumpur, 26–28 Nov, book of Abstract, pp. 77–78 (2013)
52. Yeoh, T.S., Yeoh, T.J., Song, C.L.: TRIZ: Systematic Innovation in Manufacturing. Firstfruits Publishing, Petaling Jaya (2009)
53. Wright, I.: Design Methods in Engineering and Product Design. McGraw Hill Publishing Company, Maidenhead (1998)

Chapter 6
Manufacturing Techniques of Tropical Natural Fibre Composites

Abstract In this chapter, an overview of manufacturing process for tropical natural fibre composites is presented. Among the composite manufacturing processes reported to be used to product products from tropical natural fibre composites include hand lay-up, filament winding, pultrusion, resin transfer moulding, vacuum bag moulding, compression moulding, injection moulding, vacuum assisted resin infusion and extrusion.

Keywords Hand lay-up · Pultrusion: filament winding · Injection moulding · Compression moulding

6.1 Introduction

Manufacturing processes for natural fibre composite products normally utilizing manufacturing processes to produce products from conventional fibre composites such as glass, Kevlar and carbon fibre composites. According to Ziegmann and Elsabbagh [17], the manufacturing processes that can be used to produce products from natural fibre composites include injection moulding, extrusion, pultrusion, vacuum assisted resin infusion (VARI), hand lay-up filament winding, resin transfer moulding, compression moulding. In addition, Mariatti and Abdul Khalil [13] had used vacuum bagging to produce specimens from natural fibre composites. Although the methods reported by Ziegmann and Elsabbagh [17], can be used to produce products from any types of natural fibre composites, it is believed that those methods are also applicable to tropical natural fibre composites.

Production of composite materials involves the processes of producing fibres, arranging fibres into bundles or fabrics, and introducing the fibres into the matrix to form composites [2]. Only introducing the fibres into the matrix to form composites are discussed in this chapter.

There are three ways to introduce fibre into matrix to form composites, i.e. based on manufacturing processes [14]:

© Springer Science+Business Media Singapore 2014
M.S. Salit, *Tropical Natural Fibre Composites,*
Engineering Materials, DOI 10.1007/978-981-287-155-8_6

1. Fibres and matrix are obtained separately and the user has to process them to form a composite by fabrication. The example is in hand lay-up process.
2. The user needs to have moulding compounds with the correct amount of fibres, matrix and other additives. The mouldings are placed in mould for press or injection moulding to make a component.
3. Composite components are ready to be used after fabrication process using pultrusion or filament winding. Only secondary process like assembly is required.

6.2 Hand Lay-up

Hand hay-up is a traditional process to fabricate composite products. It is also called wet lay-up process. This method is the widely used method because the tooling cost is very low compared to other methods that involved some degree of automation. Hand lay-up can produce products with excellent surface finish but it is only on one surface only, i.e. the inner surface that is contact with the mould. This method is sometimes called a contact moulding. This method generally requires skilful operators to carry out the moulding operation and therefore it is labour intensive. In this process mould has to be correctly prepared. The materials for the mould could be of wood, plaster, plastics, composites, or metal [4]. The labour cost for this process is very high too. The process involves an open mould where on the inner surface, layers of composites are placed (see Fig. 6.1). It is suitable to make large structure like air craft and sport car bodies, swimming pools, boats, storage tanks, dooms and for making prototypes [15]. In this method fibres are laid onto an inner surface of a mould. Release agent was previously applied to prevent the composites from sticking onto the mould surface and gel coat was also applied as decoration and protection of the mould surface. The process has quite long production (time consuming process).

The followings are a case study of fabrication of consumer goods product (telephone stand) from banana pseudo-stem fibre reinforced unsaturated polyester composites using hand lay-up process. Due to economic and ethical issues, banana stem used for this application was taken from the one that the fruit has been harvested, or the one that the fruit is not consumable so that the banana stem or trunk is useless. The stem was cut into a certain length, and then the stem was peeled layer by layer. The peels were then dried under the sunlight for two weeks, soaked into water for at least two weeks, and again dried. The fibres were interlaced at 90° to each other to form a woven fabric. Polymer resin and hardener/catalyst was prepared with the appropriate ratio. The release agent was laid on the mold after the mold was cleaned. First layer resin was laid up uniformly for each layer on to the mould. The woven fibers were then added into the mould. The processes were repeated until 4 layers of resin and fibres were obtained (Fig. 6.2). The composite was left for 24 h for curing. The composite was separated from the mould.

6.3 Filament Winding

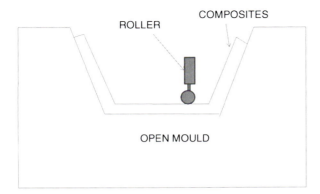

Fig. 6.1 Hand lay-up process

Fig. 6.2 Hand lay-up process to produce a telephone stand

6.3 Filament Winding

Filament winding is an automated composite manufacturing process where continuous fibres (rovings) or tapes are wound around rotating mandrel after passing through a resin bath. The types of winding of composites are possible namely in hoop, helical and polar direction (Fig. 6.3). Figures 6.4 and 6.5 show two types of resin bath used to produce composite shaft from kenaf fibres. The mandrel has the

Fig. 6.3 Plastic rope wound in polar direction

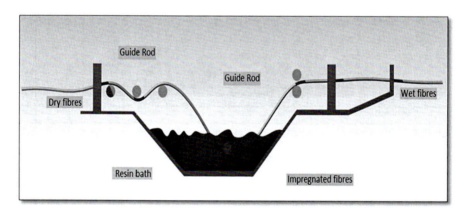

Fig. 6.4 Fibre dip type of resin bath (Courtesy of Mr. Sairizal Misri)

shape of the final product. Lilholt and Madsen [11] reported the work on flax and hemp fibre reinforced polymer composites using filament winding process. The composites are wound around the mandrel to a desired length. The mandrel can form part of the component or can be removed from the component. Resins normally used include epoxy, polyester, vinyl ester and phenolic. Fibre tensioner is used to help controlling the placement of the tow, the resin content, and layer thickness [10]. Compare to other methods, RTM can produce products with large

6.3 Filament Winding

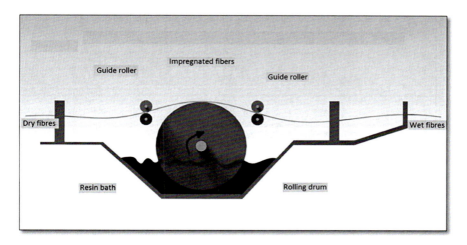

Fig. 6.5 Drum type of resin bath (Courtesy of Mr. Sairizal Misri)

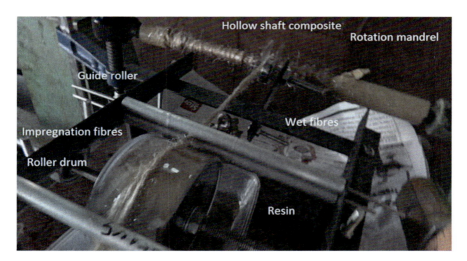

Fig. 6.6 Impregnated kenaf fibre composites in filament winding process (Courtesy of Mr. Sairizal Misri)

lengths and diameters. Circular components can conveniently made from this process such as pipes, drive shafts, pressure vessels, golf club shafts, composite oars, and fishing rod but non circular products can also be produced by this method. Figure 6.6 shows a filament winding process to produce composite shaft from kenaf fibre composites.

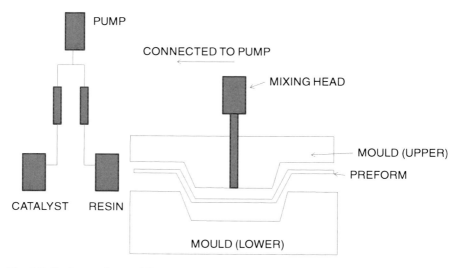

Fig. 6.7 Resin transfer moulding

6.4 Resin Transfer Moulding

Resin transfer moulding (RTM) is also called a liquid transfer moulding process [15]. It is a closed mould composite process which involves the placement of dry preform (reinforcement) inside the closed matched mould that resembles the final product. Then thermosetting resin is then injected into the mould and preimpregnated onto the surface of preform. Resin, hardener, catalyst, colour, filler, etc. are pumped into the mould using a dispensing equipment (Fig. 6.7). The resin is then cured and the composites are removed from the mould. RTM is becoming popular manufacturing method in the recent years in automotive and aerospace industries. The products produced from this method have excellent surface finish on both sides. Complex parts can be manufactured using RTM and it is a medium volume process. Fibres, as high as 65 % by volume can be incorporated in the composites. RTM can produce a part with complex shapes and large surface areas [8]. It can also produce components for sporting goods, and consumer product applications. Mazumdar [15] reported that products normally made from RTM include helmets, doors, hockey sticks, bicycle frames, windmill blades, sport car bodies, automotive panels, and aircraft components (bulkheads, space blocks, spars, fairings and control surface ribs and stiffeners.

6.5 Pultrusion

Pultrusion is a continuous process that can be used to produce natural fibre composites by placing continuous fibre rovings (Fig. 6.8) on a creel and pulling them over a resin bath or resin impregnation system that contains thermosetting

6.5 Pultrusion

Fig. 6.8 Kenaf fibre rovings (Courtesy of Innovative Pultrusion Sdn. Bhd., Seremban, Malaysia)

resins such as unsaturated polyester or epoxy (Figs. 6.9 and 6.10). The fibres leave resin bath in fully wetted out state. Then the continuous impregnated composites were pulled at constant speed by means of a puller/gripper (Fig. 6.11) through a series of custom tooling called a 'preformer' to arrange and organize the composites into the final shape and the excess resin is squeezed out [8]. Normally, mirror-polish machined steel (chromium-plated) is used for the dies [16]. Excess resin is removed so that entrapped air can be expelled and the fibres can be compacted [14]. Then the composites are pulled through a heated die with different temperature zone to cure the composites (Fig. 6.12). In some pultrusion systems, resin can also be injected into the die. The composites are then cut into lengths using a cutoff saw and finally are stacked. An example of product made from pultruded kenaf fibre composites is shown in Fig. 6.13. The matrices commonly used in pultrusion process are normally thermosetting polymers such as epoxy, polyester, phenolic and polyurethane. Thermoplastic materials can also be used in pultrusion. Mechanical properties of pultruded composites can be controlled by the fibre wet-out [12]. Pultruded components can be produced with different variety of structural shapes with lengths of constant-profile sections, either solid or hollow [14] such as I-beams, T-beams, rods, channels, angles, and

Fig. 6.9 Continuous roving travel from creels toward resin bath (Courtesy of Innovative Pultrusion Sdn. Bhd., Seremban, Malaysia)

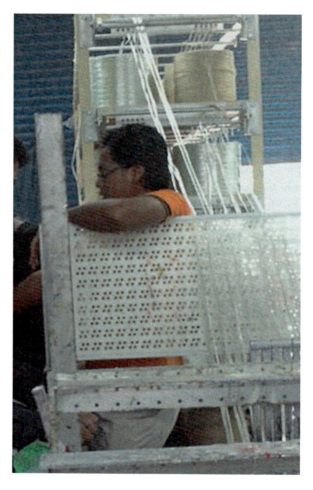

thin-walled tubes [8]. The components can be self-coloured they are difficult to join [14]. Pultrusion used low-cost fibres like glass and natural fibres and low cost resins and low-cost products can be produced.

6.6 Compression Moulding

Compression moulding is used to produce components at high production volume and it is suitable to produce automotive components [15]. It uses a heated matched mould in its operation. Thermoset bulk moulding compound (BMC) and sheet moulding compound (SMC) are two moulding materials traditionally used in compression moulding. For thermoplastic materials, glass mat thermoplastic is normally a typical moulding compound for compression moulding.

6.6 Compression Moulding

Fig. 6.10 Continuous roving travel through guides toward resin bath and forming station (Courtesy of Innovative Pultrusion Sdn. Bhd., Seremban, Malaysia)

Fig. 6.11 Gripper pulls the composite part at continuous speed (Courtesy of Innovative Pultrusion Sdn. Bhd., Seremban, Malaysia)

Compression moulding (Fig. 6.14) was first developed to manufacture composite parts to replace metals and it is typically used to make larger flat or moderately curved parts. Compression molding is a processing technique for thermoplastics and thermosets and their composites [6]. Processing thermoplastics by using compression molding is different from processing thermosets. Thermoplastics need higher heat and higher pressure to be processed because of their high

Fig. 6.12 Continuous roving travel through heated die (Courtesy of Innovative Pultrusion Sdn. Bhd., Seremban, Malaysia)

Fig. 6.13 Pultruded kenaf fibre composite product

viscosity. Compression molding starts with placing the material in an open mould. The mould is placed between two heating plates. After placing the material in the cavity of the mould it is heated to be softened for some time depending on the material. Then it is full-pressed under the required pressure to take the mould shape. After that the mould is placed between two cold plates for cooling to solidify the material before demoulding [5].

Fig. 6.14 Laboratory scale compression moulding

6.7 Injection Moulding

Injection moulding is a process of producing composites in which molten polymer and chopped fibres are injected into a mould to form composite and the composite is then cured. Injection moulding is an automated process and is used to produce products in high volume. It is a manufacturing process for making complex shapes from polymeric materials. The process involves heating up the polymer, injecting the molten polymer into a mould, cooling the polymer in the mould, and ejecting the part.

Both thermoplastic and thermosetting polymers can be used in this process. For natural fibre composite production using this method, normally injection moulding compounds (IMC) are produced. Then these compounds are fed into a heated barrel through a hopper and the granules of compounds are melted after undergoing shearing process in reciprocating screw. The melt is then injected into the mould through a sprue and runner system. In the mould, the melt enters the mould cavity through a gate. After the moulding is set, the mould is split and the component is ejected. The injection molding cycle has 4 major components, and it is indicated as follows:

- Fill time: The mold is closed at the start of the fill time. The screw moves forward, forcing molten plastic into the mould, which is normally very fast. This part of the process is controlled by velocity. As the plastic hits the cold mould wall it sticks to the wall and freezes. The plastic flow channel is located between the frozen layers of polymer. The rate of injection has a large influence on the thickness of the frozen layer. How the part is filled is the biggest single influence on the quality of the part, in most cases. Filling is normally the smallest portion of the molding cycle.
- Hold or Pack time: Once the cavity is filled, the injection molding machine continues to apply pressure to the plastic to pack more material into the mold. This is to compensate for shrinkage as the plastic continues to cool. This part of the process is controlled by pressure. The switchover from velocity control to pressure control occurs just before the part volume has been filled.
- Cooling time: There is no longer any pressure applied to the plastic. The part continues to cool and freeze until it is cool enough to eject and hold its shape. While the part is cooling, the screw rotates molten plastic for the next cycle.
- Mould open time: The mould is opened, the part is ejected, and the mould is closed in preparation for the next shot.

Figure 6.15 shows an injection moulding machine used produce products from tropical natural fibre composites.

Fig. 6.15 Injection moulding (Courtesy of Mr. Azaman Md Darus, Universiti Malaysia Perlis, Malaysia)

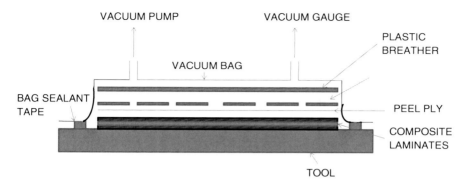

Fig. 6.16 Vacuum bag moulding

6.8 Vacuum Bag Moulding

Vacuum bag moulding (Fig. 6.16) or vacuum bagging is one of the composite manufacturing processes especially to produce products from laminated composites. It is a process where an open mould is placed with composites. Vacuum bag is used in this process where it provides compaction pressure and consolidation of laminated composites [8]. Vacuum bagging technique is in fact a variation hand lay-up process. Mariatti and Abdul Khalil [13] used vacuum bagging technique to develop bagasse fibre reinforced unsaturated polyester composites. The mould surface is covered with laminate and the next layer is perforated release film or peel ply. On top of that a bleeder is placed and the next layer is separator. Above that breather is laid. Once the lay-up is completed and placed inside a flexible film vacuum bag and all edges are sealed [16].

6.9 Extrusion

Continuous composite products made from thermoplastics and filler made from natural fibre can be made using extrusion process. The extrusion process consists of the extruder, the die, the forming stage, the post-forming or handling stage and secondary process [7]. Extrusion is a forming process to produce fixed cross sectional products such as tubes, pipes, films, sheets, profiles and strips by pushing or drawing material through die [1]. By this definition, various materials can be used in extrusion such as metal, thermoset and thermoplastic plastics, ceramic and even food. The most common method of plastic extrusion is by extruder.

Thermoplastic material needed to be heated until it is soft enough to be pushed through the die. The simplest method is by using single screw extruder (Fig. 6.17). There are many designs available for screw depending on material and the function of extruder. However, all screws have 3 basic zones or sections: feeding zone,

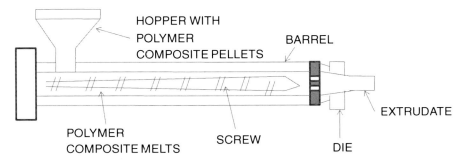

Fig. 6.17 Components and features of a single screw extruder of thermoplastic polymer and composite materials

compression (transition or melting) zone and metering zone. Feeding zone is where the material is received and heated. As the material move along the screw, the material began to soften and turned into high viscous fluid. Depend on screw design, material are mixed and/or compressed in compression zone before enter the metering zone where material are constantly pushed toward the die.

In composite extrusion, plastic pellets and fibers are compounded using compounder (such as internal mixer or extruder) into composite pellets. The composite pellets become the raw material for extrusion process and replace plastic pellets. It is possible to produce composite extrusion by combining both compounder and extruder into in line compounding and extrusion process extruder. This method rely on the design the screws where mixing zone was required to mix fiber and plastics.

6.10 Vacuum Assisted Resin Infusion (VARI)

As stated earlier Ziegmann and Elsabbagh [17] reported that vacuum assisted resign infusion (VARI) method has been used to produce products from natural fibre composites (Fig. 6.18). Unlike wet hand lay-up, the operator does not have to roll resin into the fibres but instead the dry fibre or prefrom is placed inside the mould and resin supply tubes and vacuum plumbing is placed on the backside of the lay-up [8].

In the field of composites, resin infusion is a process where the voids in an evacuated stack of porous material are filled with a liquid resin. When the resin solidifies, the solid resin matrix binds the assembly of materials into a unified rigid composite. The reinforcement can be any porous material compatible with the resin. Typical materials are inorganic fibres (with glass fibre being most common), organic fibres such as flax, kenaf and etc., or combinations of fibres with other materials such as closed cell foams, balsawood, and honeycomb. Porous materials can also be infused onto the surfaces of non-porous materials such as sheet metal.

6.10 Vacuum Assisted Resin Infusion (VARI)

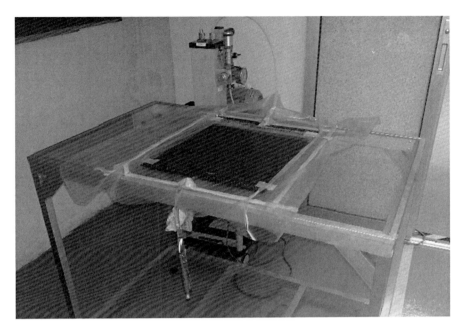

Fig. 6.18 Vacuum assisted resin infusion (VARI) process (Courtesy of Mr Azaman Md Deros, Universiti Malaysia, Perlis, Malaysia)

Resins are usually thermosetting types, but thermoplastic resins can also be used for infusion. A key part of the process is the evacuation, or removal, of the air from the porous material prior to admitting the resin. The air needs to be removed from the porous material to allow the resin to take its place [4]. The vacuum assisted resin infusion process is a cost effective method of manufacturing high quality and high strength composite parts that are required in relatively low quantities, say less than a few hundred identical pieces per mould per year, or physically large parts which are difficult, or prohibitively expensive to make by any other method (Azaman 2014).

References

1. Anon.: http://en.wikipedia.org/wiki/Extrusion (2014). Accessed 2 Feb 2014
2. Askeland, D.R., Fulay, P.P.: Essentials of Materials Science and Engineering, 2nd edn. Cengage Learning, Toronto (2009)
3. Azaman, M.D. (2014), Personal Communication, Universiti Malaysia Perlis
4. Barbero, E.J.: Introduction to Composites Design. Taylor & Francis Inc, Philadelphia (1999)
5. Biron, M.: Thermoplastics and Thermoplastic Composites: Technical Information for Plastics Users. Elsevier, Oxford (2007)
6. Chanda, M., Roy, S.K.: Plastics Technology Handbook, 4th edn. CRC Press, Boca Raton (2007)

7. Clegg, D.W., Collyer, A.A.: The Structure and Properties of Polymeric Materials. The Institute of Materials, London (1993)
8. Dorworth, L.C., Gardiner, G.L., Mellema, G.M.: Essentials of Advanced Composite Fabrication and Repair. Aviation Supplies & Academics Inc, Newcastle (2009)
9. Herakovich, C.T.: Mechanics of Fibrous Composites. Wiley, New York (1998)
10. Hyer, M.W.: Stress Analysis of Fiber-Reinforced Composite Materials. WCB McGraw-Hill, Boston (1998)
11. Lilholt, H., Madsen, B.: Properties of flax and hemp composites. In: Reux, F.,Verpoest, I. (eds.) Flax and Hemp Fibres: A Natural Solution the Composite Industry, pp. 119–140. JEC Composites, Paris (2012)
12. Mallick, P.K.: Fiber-Reinforced Composites: Materials, Manufacturing, and Design. CRC Press, Boca Raton (2008)
13. Mariatti, J., Abdul Khalil, S.: Properties of bagasse fibre-reinforced unsaturated polyester (USP) composites. In: Salit, M.S. (ed.) Research on Natural Fibre Reinforced Polymer Composites, pp. 63–83. UPM Press, Serdang (2009)
14. Mayer, R.M.: Design with Reinforced Plastics: A Guide for Engineers and Designers. Design Council, London (1993)
15. Mazumdar, S.K.: Composites Manufacturing: Materials, Product, and Process Engineering. CRC Press, Boca Raton (2002)
16. Murphy, J.: Reinforced Plastics Handbook. Elsevier, Advanced Technology, Oxford (1994)
17. Ziegmann, G., Elsabbagh, A.: Production techniques for natural fibre composites. In: Reux, F., Verpoest, I. (eds.) Flax and Hemp Fibres: A Natural Solution the Composite Industry, pp. 99–117. JEC Composites, Paris (2012)

Chapter 7
Applications of Tropical Natural Fibre Composites

Abstract This short chapter presents the application of tropical natural fibre composites. It started with the use of these materials in automotive components mainly in Malaysian context. Academic exercises of the author and his colleagues in the form of products are also presented.

Keywords Products · Natural fibre composite components · Composite furniture · Kenaf · Composite fabrication

7.1 Introduction

Tropical natural fibre composites have great potential to replace many conventional fibre composites in numerous applications ranging from furniture, automotive components and cutleries. In this chapter, selected applications of tropical natural fibre composites are presented.

7.2 Application

Gomina [1] stated the use of 'green' materials in automobiles can have positive impact on the environment in terms of reducing CO_2 emissions and in turn reducing the environmental footprint in automotive industry. Westman et al. [7] reported that automotive manufacturers such as Mercedes Benz, Toyota and Daimler Chrysler had used natural fibre composites in interior and exterior parts in automobiles. Weight reduction (fuel efficiency) and sustainability of manufacturing process are two major drives in using these materials. According to Luo and Netravali [3], PALF composites had been used as materials for parts of Mercedes-Benz commercial vehicles in Brazil.

© Springer Science+Business Media Singapore 2014
M.S. Salit, *Tropical Natural Fibre Composites*,
Engineering Materials, DOI 10.1007/978-981-287-155-8_7

Fig. 7.1 Drain cover made from pultruded kenaf composites (Courtesy of Innovative Pultrusion Sdn., Bhd., Seremban, Malaysia)

In Malaysia, Polycomposite Sdn. Bhd. based in Kajang, Selangor produced natural fibre composite with the trade name of POLYMEX for automotive components such as automotive interior trims (package tray, door panel and speaker shelf) for national and international cars such as PROTON, HONDA, HYUNDAI, NAZA (KIA), TOYOTA and PERODUA. The company used rice husk, oil palm fibres, rubber and wood as fibres in polymer composites to produced engineering products [5].

A project was funded by Malaysian government on the use of kenaf fibre composites in automotive industry [2]. Among the activities carried out include acenomic aspect of kenaf composites for automotive industry, preparation and characterization kenaf based composites for automotive components, machinability and mouldability of kenaf composites for automotive components. Malnati [4] reported that coir fibre (coconut husk) polypropylene composite has been used by Ford in trunk liner.

A company in Malaysia, Innovative Pultrusion Sdn. Bhd. has developed drain cover from pultruded kenaf fibre reinforced epoxy composites (Fig. 7.1). The advantages of this product include light weight, more robust than metal counterpart, available in attractive colour without the need to paint, enhance the beauty of a house, cheaper, minimum maintenance, no electric shock, not affected by white ant, non corrosive, good against sunlight and save from theft [6]. In Malaysia drain cover made from steel is normally stolen by thieve because of the price.

Fig. 7.2 Pultruded rod from kenaf composites

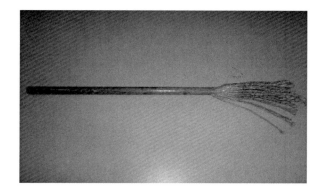

7.3 Miscellaneous Products Developed by the Author

Figures 7.2, 7.3, 7.4, 7.5, 7.6 and 7.7 show the products made from tropical natural fibre composites by the author and his colleagues. They include pultruded rod, chair, telephone stand, table, trolley base and small boat made from various tropical natural fibre composites such as kenaf, banana pseudo-stem, and sugar

Fig. 7.3 Chair from banana stem composites

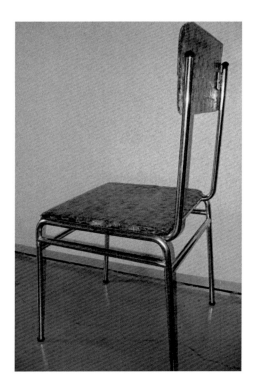

Fig. 7.4 Telephone stand from banana stem composites

Fig. 7.5 Table from banana stem fibre composites

7.3 Miscellaneous Products Developed by the Author

Fig. 7.6 Trolley base from kenaf fibre composites

Fig. 7.7 Small boat from hybrid glass-sugar palm composites

palm reinforced composites. The matrices used are mainly thermosetting polymers such as epoxy and unsaturated polyester. The manufacturing processes include hand lay-up and pultrusion. These products are developed in the form of prototypes and the main purpose is just for academic exercises. However, there is a huge potential to upgrade the products so that they can later on be commercialized.

References

1. Gomina, M.: Flax and hemp composite applications. In Reux, F.,Verpoest, I. (eds.) Flax and Hemp Fibres: A Natural Solution for the Composite Industry, pp. 141–162. JEC Composites, Paris (2012)
2. Ishak, Z.A.M.: Natural fibers as sustainable and eco-friendly materials for lightweight automotive composites, presented at National Automotive Technology and Digital Engineering Symposium 2012, May 14–15, Shah Alam, Selangor, Malaysia (2012)
3. Luo, S., Netravali, A.N.: Mechanical and thermal properties of environmental-friendly "green" composites made from pineapple leaf fibres and poly(hydrobutyrate-co-valarate) resin. Polym. Compos. **20**, 367–378 (1999)

4. Malnati, P.: Natural fiber composites drive automotive sustainability. Composite Technology **18**, 46–48 (2013)
5. Polycomposite Sdn. Bhd.: Established manufacturer of natural fibre composite (2012). http://www.polycomposite-sb.com/index.html
6. Senawi, R.: Innovative Pultrusion Sdn. Bhd., Seremban, Malaysia, personal communication (2012)
7. Westman, M.P., Laddha, S.G., Fifield, L.S., Kafentzis, T.A., Simmons, K.L.: Natural Fiber Composites: A Review, report no. PNNL-19220. Report prepared for U.S. Department of Energy (2010)